talisman

*It is not the critic who counts, not the man who points out how the strong man stumbled or where the doer of deeds could have done better. The credit belongs to the man who is actually in the arena; whose face is marred by dust and sweat and blood; who strives valiantly; who errs and comes short again and again . . . who knows the great enthusiasms, the great devotions, and spends himself in a worthy cause; who at the least knows in the end the triumph of high achievement; and who, at the worst, if he fails, at least fails while doing greatly, so that his place shall never be with those cold and timid souls who know neither victory nor defeat.*

– Teddy Roosevelt

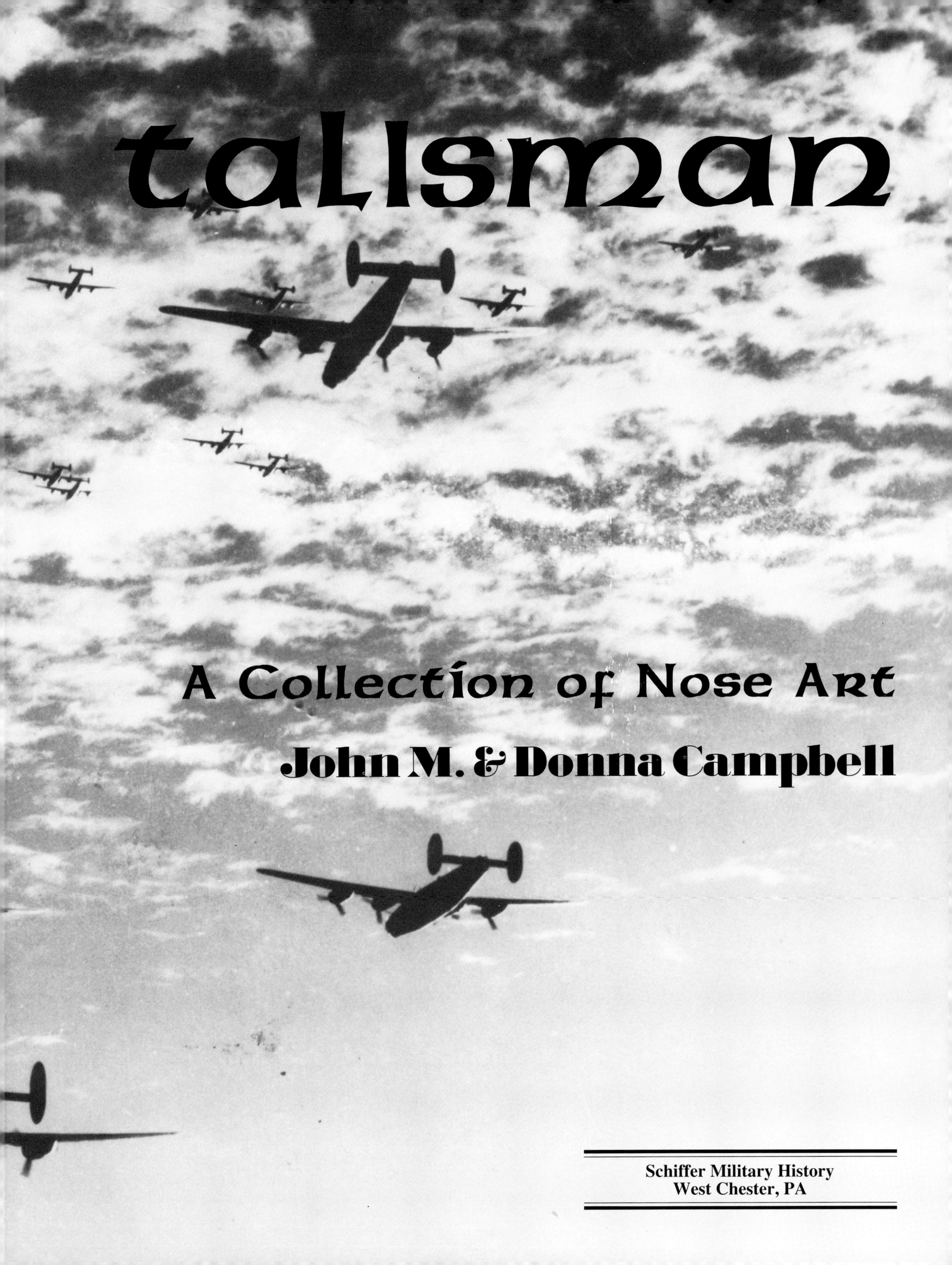

# talisman

## A Collection of Nose Art

John M. & Donna Campbell

Schiffer Military History
West Chester, PA

*Front Dust Jacket:*
After six weeks of radar school, Don Acord and Bob Klief joined the 530th Bomb Squadron, 380th Bomb Group on the Island of Mindoro, in the Philippines, where this new radar equipped B-24M-5-FO, serial 44-50602, complete with art and name was waiting. *DOUBLE TROUBLE*, so named because of the double duty the aircraft was capable of handling, flew lead ship with Acord and Klief alternating missions. They soon picked up the title *Mickey* operators because they were in charge of the radar equipment nicknamed by the group *Mickey Mouse* equipment. The radar had over 32 buttons and a radar dish housed in a dome that was manually cranked down after reaching cruising altitude. *Acord*

*Rear Dust Jacket:*
*I'll Be Seeing You*, a B-24J, belonging to the 529th Bomb Squadron, 380th Bomb Group. Named for a popular song of the times, *I'll be Seeing You*, pictures memories of that last kiss before leaving while avoiding the words *good-bye*. The sticker on the suitcase shows the destination to be the Southwest Pacific. *Acord*

Book Design by Robert Biondi


Library of Congress Catalog Number: 92-60358



Printed in the United States of America.
ISBN: 0-88740-414-6

We are interested in hearing from authors with book ideas on related topics.

Published by Schiffer Publishing Ltd.
1469 Morstein Rd.
West Chester, PA 19380
Please write for a free catalog.
This book may be purchased from the publisher.
Please include $2.95 postage.
Try your bookstore first.

# Contents

# Acknowledgements

An extreme amount of assistance is required on a project covering as vast a subject as nose art. The best part of this is the wonderful group of people involved in the perpetuation of aviation history which are so willing to help. There are no words to express the feelings we have for those who have so willingly given of themselves and their time to help us with this pictorial history. In our research we have corresponded with pilots, crew members, ground crewmen, or sometimes the family they left behind. We have received assistance from writers, historians and just plain wonderful people without whose assistance *Talisman* would not exist.

Our special affection and appreciation must go to our good friend and mentor, Jack Moses, for his inspiration and support in this and other aviation projects. His service to the Oklahoma Air Space Museum and Oklahoma Aviation is an tribute to his inherent love of the old flying machines, and his unquestioned devotion to aviation history.

Our thanks to Dr. James H. Kitchens III, Archivist, History Dept. Maxwell AFB as well as our friend and fellow photographer, Al Dodson, for their assistance in our acquisition of photos from that facility. Also Charles C. Worman, Head of Research Division, and Wes Henry Researcher, Air Force Museum, Wright Patterson AFB.

Our sincere personal thanks to the staff of the Tinker AFB Office of History, Dr James Crowder, Dr. Donald Klinko, and Lura M. Casey, for the hours of research they saved us. Also the Oklahoma Air and Space Museum, Kirkpatrick Center OKC, Museum Director Don Finch, and his Executive Assistant, Stewart Howard, for the use of the OASM Library.

We wish to extend a special word of thanks to Frank and Frieda Sanders, of the 79th Fighter Group, for all of their efforts in securing the fine photo albums from North Africa including the Dr. Charles Jaslow collection, as well as that of Dr. Hoffman. We wish to congratulate them and others for having the foresight to take photos in a time when film was so rare.

We wish also to thank our good friend J.R. Beitling and his wife Carol for inviting us into their home and assisting us in the identification of the endless B-24 photos. Also Mike Hill for his assistance with the 98th Bomb Group and 451st Bomb Group Ploesti mission extracts, Herman Hetzel with the 458th Bomb Group, and Glenn Horton and Arden Acord with the 380th Bomb Group. Maj. S.D. Huff of the 49th Fighter Squadron, 14th Fighter Group for his continued support. Also Roy J. Hogue of the 79th Fighter Group for the use of his negatives. Emil Mesco for his assistance with B-29s from the 44th Bomb group. Robert Hernandez for his contribution of B-26 photos and John Shaw for the P-38s from the 8th Fighter Group. Laxton and Frieda Malcolm for their contribution of photos. Our thanks to fellow writer and friend Jeff L. Ethell for his assistance with photos which helped make this project a success. Our thanks to Ed Janesic, Victor Veroda, Merle Olmstead, and the late Ed Haskamp for allowing us to use their photos. Special thanks must be extended to A. Kevin Grantham, Mark Copeland, Julian Myers, Robin Hubbard, Jack McKee, Jerry McKee, Ed Kueppers, James Crow, Mike Moffitt and William Cleveland. These fine men all assisted in finding some of the most sought after materials and secured the loan of these to our archives. Ron Witt for the loan of many of his negatives. Paul M Andrews, and William Adams for their research into heavy bomber losses and aircraft names. Also Larry Davis, Sgt. Mark Bacon, Gerald Cullumber, Grant Delong.

Our deep appreciation goes to the P-38 National Association for their continued enthusiasm and support of our work and to our very special friends Col. Joseph A. Kuhn and Jack M. Ilfry who have both been of great assistance and will always be held in our highest regard. A.T. Lloyd for his assistance with the Boeing files. Special thanks to Robert C.Mikesh for the use of his wonderful slides. Vyron DeHas, Stanley Sommers, Steve Sheflin, William Swisher. We also appreciate the assistance and technical advice we received from Stew Ostler of Epperson Photo and Ron Kirkwood, Tonia Crawford, Bryan Wood, Chuck Blackwell and Rick Henn of Kirkwood Slide Services. Lynn Loftin, Brenda Smith, Yvonne Cagle, Stephanie Proctor, Roy & Robyn Ikeda of Moto Photo for their excellence in color photos and our good friend Joann Essig for her gracious assistance in her color lab. Randy Walker for his assistance.

Our special thanks to Marvin Kinchen, "Computer Wizard" for his interest in seeing the logistics of these projects work to their maximum. Thanks to our very close friends Sandra Breed and Theresa Privette for their support. To our California connection, Nick Veronico, for his assistance in getting the word out. Garry R. Pape for securing color from the SWP. Ruth Havens, Garland Leonard, Jon Maguire, Mike Conners, Gary James, Wayne Waldrond, Wayne "Donnie" Watts, Arnold Fort, Nancy Heinbach, Eulene Harbor, Bob Young, Tony Chappel, Claude Cranfield, Gene Bertwell, Rob Hubbard, Mike Green, Jay Reid, Bill Hess, Jerry Nevells, Fenton Morrison, Henry Fennel, Judy Willoughby, Steve Link, Bob Klief, Stanley Davis and Earl Kielgas for their contributions.

We in closing our acknowledgements would be most remiss indeed if we didn't extend warm thanks to our parents F.D. and Ruth Campbell and Jess and Jewell Easton for their continued support throughout our lives.

*To Jackie:*
*Who brought sunshine into our lives.*

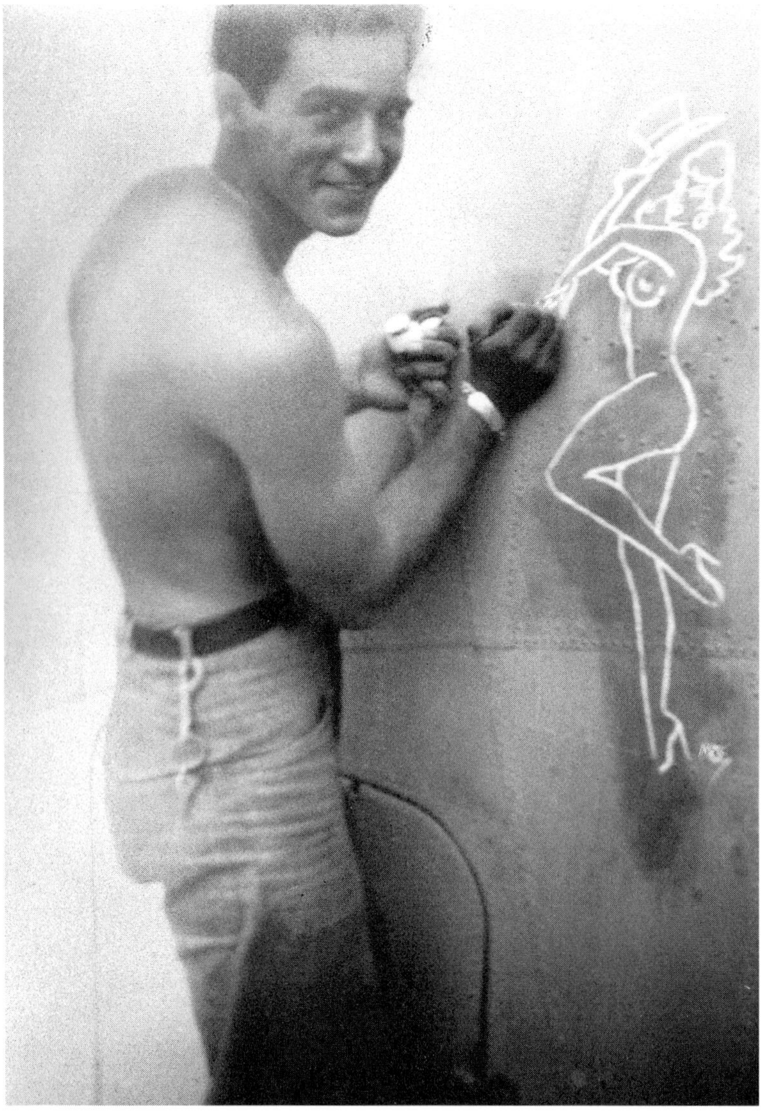

Jack Moses, when on duty, was a gunner on such aircraft as the Ventura and Mariner aircraft in the U.S. Navy. When at play he was a graphic artist. The art, *Panama Hattie*, designed and named by Jack, is being painted on a Mariner. Jack, being a clever artist, positioned the art to take advantage of the planes rivets on the breasts. *Moses*

THESE PHOTOS DEPICT THE DILIGENT ARTIST AT WORK.

Jack Moses is best known nationally for his pen and ink drawings of the valiant years of World War I in the air, as seen in *Over The Front* magazine. Locally he is known for his success and artistic talent as a commercial artist. Jack has been able to use this talent to keep him close to his first love, aviation, as he has been commissioned to do paintings of military aircraft, one of which hangs in the Fighter Aces Museum located in Mesa, Arizona. Aviators in the state of Oklahoma know him best by his incredible collection of aviation militaria displayed at the Oklahoma Air and Space Museum at the Kirkpatrick Center, of which he is not only a large donor but also holds the title of Assistant Curator. Among his more unique honors, Jack is also one of the very few, honorary members of the famed Lafayette Escadrille Flying Corp. As a friend, Jack is known and respected by many as a "Gentleman's Gentleman."

*The Shady Lady* was a PBY-5A Catalina flying boat assigned to Headron at N.A.S. Panama. This aircraft entered service in 1942 and flies the flags of sixteen foreign countries. *Moses*

*Casa de Hot Rocks* was the name given this PBM5-E Mariner assigned to VP-74. This aircraft anchored off the Galapagos illustrates that, while nose art in the U.S. Navy was not common, slogans did in fact exist. *Moses*

# Foreword

# Remembrances

People have long discussed the phenomenon of aircraft decoration, referring to it as a "Talisman", or lucky sign. Some associate it with a personal icon that, we can argue, goes back to the prehistoric ritual of painted hunting scenes on walls of the "Grotte de Lascaux", near what is now Montignac, France, or the cavern of La Greze, also in France, or an engraved walrus tusk from Lapland in northern Scandinavia. All of these executed in an effort to gain favor from the Gods (or Goddesses) assuring a successful hunt.

Early man struggled through an inexpressibly dark chaos of fear and violence, and relied on tools, weapons and primitive religion for protection and survival. During the long journey since, modern man has still evinced the same emotions, if not the same spectres, with his tools, weapons and psyche that only equips him to understand his relationship to some Supreme Being.

Our mind's eye abounds with visions of Egyptian chariots, my own ancestor's war galleys, of heraldic knighthood, decorative figureheads on the prows of sailing ships, and the sailor's tattooed bodies. This ritual later incorporated a warrior's personal mount, as was the case with General Robert E. Lee's "Traveler" who would share fame with his master. A powerful weapon might be named "Old Betsy" after Daniel Boone's rifle, and become famous in hunting folklore. Indian war parties marking their horses and themselves in war paint as a charm, and in supplication to the Great Spirit.

In peacetime we adorned our bikes, motorcycles, Chevys, Fords and other jalopies with foxtails, reflector-studded mud flaps, radiator ornaments and even some forms of lettering on the car's chassis itself, a la "Andy Hardy." When these young men entered military service, they spontaneously transferred this same desire of self-expression to the inanimate objects in their charge – cannon, jeep, truck, tank or airplane – to differentiate it from all the other look-alikes.

Fortunately, the military generally looked on this action with patronizing approval, regarding it as a morale builder, though its acceptance varied greatly within different branches of the service.

In their first nose art book, "*WARPAINT*", John and Donna Campbell set a high standard for this genre of artwork as they gathered and annotated a vast, broad and valuable array of nose art photos that transport us back to another world – a world that hopefully never will be or can be again; a world the existed in my own lifetime, that now seems like another life, as we might ask ourselves, "was I really ever that young?", or "Did this really happen to me?"

We now have a second offering from them. Their latest chronicle is a fascinating examination of nose art, patiently culled from decades of collecting from cooperative individuals, private collections, museums, and archives. This latest assortment forms a rich reference source of illustrations and lettering in an application of bon motts, metaphors and double entendres in witty, pithy, whimsical, sarcastic, prophetic, outrageous, but rarely religious messages. Names and slogans sometimes endear to all but the crews of the planes themselves, or to more astute GI's.

I am grateful for their obsession, and for the unearthing of more of these poignant, visual remains of the golden age of nose art. Every page shows the reader through the eye of the camera, a view of some of the players and props from one of the greatest dramas in history – and the many faces of war are brought into focus.

If beauty is in the eye of the beholder, so is history – these photos are history.

Enjoy!

Jack Moses

# Introduction

The dictionary defines Talisman as, "Something that produces extraordinary effects, especially in averting evil . . . an object believed to give supernatural powers of protection to its bearer." A study of any society will reveal a belief in some form of talisman, and its use is not confined to any place, people, or time. During the times man has felt unprotected in this world, he has sought out the sense of security it could afford. The origin of talismans can be traced to prehistory when early man adorned himself for battle seeking the favor of the gods. In the

Twentieth Century the talisman took a new and unconventional form . . . nose art.

This time the vehicles of war were being adorned, not the warriors themselves. Nose art employed a myriad of styles and themes, because the inspiration for each came from within the individual warrior and sprang from diverse motivations. To some, nose art served as a battle cry or a threat to harass the enemy. To others, it was reminiscent of their home town, the ladies left behind, and the "good ol' USA." On the sides of the aircraft you might find cartoon characters, favorite pinups, or the name of a girlfriend. There were also menacing shark mouths and

*Below:* The art which appeared on the aircraft also found its way on almost any surface that would hold paint. These voluptuous ladies were the art of Sgt. Joe Pumphrey of the 79th Fighter Group. The "Desert Rats" of North Africa had little but sand to look at outside, but once inside the club house this mural of ladies would tend to lead even the most stressed of minds to a better thought, a better time, or better things to dream about. *Hogue via Sanders*

degrading illustrations of the enemy. In reference to the game of chance the airmen were confronted with, many aircraft carried gambling terms familiar to both dice and card games and were accompanied by illustrations. Some of the art expressed the crew's feelings towards their aircraft. One P-61 belonging to the 422nd Night Fighter Squadron carried the name *The Creep*, which was later changed to *For Sale Cheap*. The P-51 piloted by Capt. Richard Peterson of the 364th Fighter Squadron, had the words *"HURRY HOME HONEY"* painted on the nose. These were the words used by his wife when closing her letters to him.

As you browse through the nose art from the various war fronts, you will find recurring pattern. You will also observe that although nose art took many forms, the one constant was the appearance and reappearance of the female figure. There were ladies! They were pictured as redheads, blondes, and brunettes. They surfaced in all manor of dress and undress. They appeared on the nose of the plane, the rudder, and anywhere in between . . . not excluding wheel covers and inside landing gear doors. Some were named for sweethearts, some for famous actresses or songs of the time, and others after comic strip characters. It was unrestrained sport, but there was an underlying somberness and urgency about both the time and the art. These young men were the warriors of the day, and like their ancestors, their art served as a talisman or shield. Their talisman provided them with supernatural protective powers during long and dangerous missions, and would afford them that extra edge against the enemy to complete the mission and return the victor. The ritual of a quick pat or a thrown kiss to a voluptuous female figure on the side of the plane was routine before take-off.

The importance of the art was not limited to crew members. Once a name, character, or figure of a girl was applied to any part of an aircraft, it was no longer solely identified by a set of numbers. Henceforth, it had its own identity, a name that recalled a vivid picture of the planes history and its crew. No memories are recalled when you hear B-24D serial #42-40081. Yet, once mentioned, the name *Lady Be Good* produces a vision of the majestic B-24 lying crumpled in the Libyan sand where she crash-landed in April of 1943. You feel the anguish experienced by the friends and family of the five brave crew members as they pondered the mystery that had surrounded the loss of both the aircraft and her crew.

The exploits of the crews and aircraft made many of the names given to the aircraft famous. Among bombers, the names *Enola Gay* and *Memphis Belle* are still remembered although it is nearing fifty years since they last flew. The B-17 *Memphis Belle* is remembered as the first bomber to complete 25 operational missions while the *Enola Gay*, the best known B-29 Super Fortress, was used to drop the first atomic bomb.

This look over the shoulder of a World War II artist shows all of the needed ingredients for that period's art. The artist is equipped with cans of paint and brushes. Stuck to the truck is a favorite pin-up of the time used as an example. The truck receiving the beautiful art belonged to the 79th Fighter Group. *Hogue via Sanders*

Bombers provided the artist with a very large canvas, and many of the phenomenal pieces of art took advantage of the entire surface. A smaller canvas was found on the bomber's "*little friends*" – the fighters, although much smaller, displayed as many varied and distinctive names and characters as did their counterparts. The highest scoring American fighter pilot, Maj. Richard Bong, known as *"The Ace of Aces,"* immortalized his wife's name *Marge* by painting it on his P-38. Among other names made famous by top scoring aces were *SHORT-FUSE SALLE* flown by Maj. Richard Turner and *PETIE* flown by Lt.Col. John Meyer.

Although they received most of the attention, the bombers and fighters were not alone in the air, nor were they alone in their flare for nose art. Nose art was applied to photo recon, transport, liaison and observation aircraft. If it flew, it was susceptible to nose art.

The same art was also applied to land and water vehicles, club houses, tents, flight jackets, and numerous other objects, but the name nose art was uniquely associated with aircraft.

Nose art, whether it was applied for luck, for fun, or for love, stands as a tribute, not only to the ingenuity of free men but to their love of home, family, and everything that is American. Although *TALISMAN* is a collection of nose art seen mostly during World War II and Korea, we as lovers of that art form, are pleased to see it resurfacing with the same degree of imagination on current day aircraft. As historians, we urge that these photos and their history be preserved for the future. President George Bush said it best when he said, *"We owe it to our future not to forget our past."* It is in this spirit that we present this study – along with a bit of nostalgia.

All of the might of the squadrons belonging to the 38th Bomb Group came together to fight the Japanese as shown on the B-25, *HARDSHIPS 2nd*. Pictured is the American Eagle perched on the Rising Sun with personalized bombs approaching their target. Each of the bombs carried a 38th Bomb Group squadron insignia. The panther belonged to the 822nd Squadron. The tiger head was the 823rds insignia. The wolf head was on the 71st Squadron's aircraft. The green dragon was on the noses of the 405th Bomb Squadron. *AFM*

## Chapter One

# Medium Bombers and Cargo Aircraft

Whether you flew in B-25 Mitchells, B-26 Marauders, A-26 Invaders, A-20 Havocs, B-34 Hudson/Venturas or the cargo C-47 Skytrains, one thing is most apparent – you are in a twin engine aircraft with a smaller crew, smaller bombload or cargo, and shorter range than most of your contemporaries. This in no way diminishes the importance of the role played by the mediums during the war years.

The first benchmark made by a medium bomber during World War II, was by the "Doolittle Raiders" in B-25s which were launched from the U.S.S. Hornet to attack Tokyo just a few short months after the Japanese attack on Pearl Harbor.

B-26 Marauders were infamous for the bombings of Monte Casino, Florence, Belgium, Holland, and many other targets ranging from France, to North Africa, and including fortress Germany. The Marauder was affectionately referred to as "The Flying Prostitute" by the men who flew her claiming she had no visible means of support. Despite this she still maintained an impressive record of target missions.

The A-20 had a legacy of her own, from its use primarily in the Southwest Pacific for bombing Japanese bases and shipping, to its conversion to the P-70, radar equipped night fighters, which served as the predecessor to the famed P-61 "Black Widow."

The Lockheed B-34, Hudson, was used both as a medium bomber and cargo transport by the Army Air Corps. In the U.S. Navy's designation, the Ventura, it was used for Costal Command's anti-submarine warfare patrol as well as the air taxi role.

The C-47, Skytrain, proved on many occasions to be the pride of the Douglas Aircraft Company. Designed as a troop transport, the Skytrain transported troops and cargo, but was also called on to tow gliders on the D-Day invasion of Normandy during the wet and overcast conditions. This aircraft alone stood out among many others simply because the C-47s were built to last. In fact, the C-47's service spans a period covering World War II, Korea, and the Viet-Nam conflict. During the latter of which, it was often referred to as "Spooky" and carried Vulcan mini guns to weed out belligerent North Vietnamese troops hiding in the dense jungle undergrowth.

The A-26, Invader, also built by Douglas, saw action in all three arenas. During the Korean Conflict it gained fame as a night bomber. After the conclusion of the war it was used as a borate fire bomber in the Pacific Northwest by the Forestry Department.

# A-20's

*"We Deliver Daily"* was a 16th Observation Squadron A-20. *OASM via Moses*

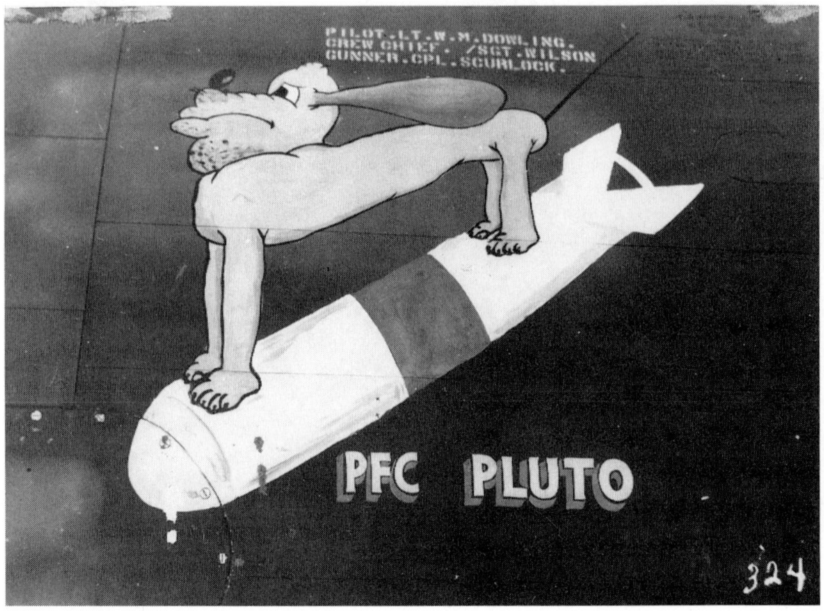


*PFC PLUTO* , an A-20, was piloted by Lt. W.M. Dowling of the 16th Observation Squadron. *OASM via Moses*

*DINAH MITE*, an A-20, was piloted by Lt. J.R. Delapp of the 16th Observation Squadron. *OASM via Moses*

The A-20, *GYPSY ROSE LEE/WE OUTSTRIP EM' ALL*, was named after the most famous burlesque queen of the time. This aircraft flew with the 16th Observation Squadron. *OASM via Moses*

The art on this A-20 belonging to the 16th Observation Squadron features the American Bald Eagle, as the hunter it is, poised on a bomb. *OASM via Moses*

*LITTLE CHIEF*, an A-20 which served as the squadron hack for the 90th Bomb Group, sports a female wearing a blouse, and shows the skull and cross-bombs of that group. She is reclining on a parachute with her feet propped up on cases of gin and rations. *Smith*

*Satan's Daughter*, an A-20 serving with the 410th Bomb Group, shows seventy bomb missions completed at the time it visited Base Air Depot #2. The aircraft was piloted by Lt. R.E. Ott. *Campbell*

*La France Libre*, an A-20, carries both the American and French flags. This 410th Bomb Group aircraft had completed 100 missions when it had its photo taken at Base Air Depot #2. *Campbell*

*THE "GREEN" HORNET*, an A-20G-35-DO, serial 43-10211, was assigned to the 410th Bomb Group. At the time of its visit to Base Air Depot #2, this aircraft had completed fifty-eight bomb missions. *Campbell*

This 410th Bomb Group A-20 has two separate pieces of art. The first and largest is of a young girl swinging. Closer to the nose of the plane is a small reclining female. This photo was taken at Air Base Depot #2, Watton, England. *Campbell*

*STAND CLEAR PLEASE*, an A-20, was piloted by Capt. T.A. Shea of the 16th Observation Squadron in North Africa. *OASM via Moses*

*Fertile Myrtle*, a 416th Bomb Group A-20, was piloted by Capt. Prentise. Its bombardier was Lt. R.T. McBrien. This photo was taken a Base Air Depot #2, Watton. *Campbell*

*Spirit Of The Order Of The Golden Chain*, an A-20 belonging to the 416th Bomb Group, served with the Ninth Air Force. This aircraft is pictured at Base Air Depot #2. *Campbell*

This carrot toting Bugs Bunny was seen on the A-20, *IDIOT'S DELIGHT*, assigned to the 9th Air Force, in the ETO. *Stegner*

Seen standing on her nose after a landing mishap is the A-20, *Norma*. *Stegner*

*MIS-A-SIP*, an A-20, serial 43-21429, was piloted by Lt. Hughes. *MAFB*

The A-20, *Eloise*, has an unique nose art using the skull and cross bones designed around the four fixed forward guns. *Watts*

The art on the A-20, *OKLAHOMA BELLE*, used the familiar *Pistol Packin Mama* female. Although the art was familiar, the sixty-two bomb markers were certainly painted in a unique pattern. *Crow*

# B-25's

*Pin-up Girl*, a B-25, was assigned to the 12th Bomb Group. Note the effects of the .50 caliber gun blasts on the side of the aircraft. At the time this photo was taken *Pin-up Girl* had been on forty missions. *Alt*

*"Calcutta Commando,"* named for the local community, was one of the B-25s belonging to the 82nd Bomb Squadron, 12th Bomb Group. *Alt*

*JIVIN' JULIE*, a B-25, was assigned to the 82nd Bomb Squadron, 12th Bomb Group. *Alt*

In this photo of the B-25, *PROBLEM CHILD*, you see the bull dog common in the art found on many of the aircraft of the 12th Bomb Group. *Alt*

The 12th Bomb Group also had its *DOUBLE TROUBLE*, as seen on this B-25. Note also the art on the nose which many of their aircraft had. *Alt*

This 100 mission B-25, *PAPER DOLL*, was one of the 12th Bomb Group's aircraft. *Alt*

Obviously the name on this B-25, belonging to the 12th Bomb Group, has been cleaned up somewhat. The "double entendre" is still there but not as strong in the new name, *EATIN' KITTY*. *Alt*

The art on this B-25, belonging to the 12th Bomb Group, is the work of L. Babich. *"GO'IN MY WAY" ?*, was the name of a movie starring Bing Crosby. The aircraft had been on fifty-five missions at the time of the photo. *Alt*

*VIKIN'S VICIOUS VIRGIN*, a B-25, was assigned to the 82nd Bomb Squadron, 12th Bomb Group. *Alt*

*"Sweet" Lorraine* is adorned with a popular pin-up of the time. It is on a 12th Bomb Group B-25. *Alt*

*Hettie's PRIDE AND JOY!* was a B-25D-20-NA assigned to the 83rd Bomb Squadron, 12th Bomb Group. *Hernandez*

*Silver Streak* was the name given to this B-25 belonging to the 12th Bomb Group. *Alt*

*KUNAI KUTTER* was the name given to this tiger head B-25G belonging to the 823rd Bomb Squadron, 38th Bomb Group. *Malcom*

The B-25, *"LITTLE JOE"*, was quite successful against enemy shipping as shown by the seven ships included in its mission marks. *LITTLE JOE* was assigned to the 22nd Bomb Group, Fifth Air Force, serving in the Southwest Pacific until V-J Day. *Smith*

It was a dirty job, but somebody had to do it. On *PROP WASH* the honor went to our little corporal seen on several of the 12th Bomb Group's B-25s. *Alt*

In this image of a 12th Bomb Group B-25 we see the business end of one of the practitioner's of the Burma Dental Clinic, "Bridge busting, Our specialty." We see four gruesome .50 caliber machine guns and one extremely lethal 75mm cannon for busting bridges and ground targets. This armament coupled with the advantage of the parafrag bombs proved to be quite the match for the heavily fortified Japanese defenses. *Alt*

Carrying the death's head over pilot's wings across the entire nose of the plane is this 490th Bomb Squadron's B-25J. This squadron assigned to the 341st Bomb Group was known as the "Burma Bridge Busters." *Maguire*

This B-25, *MONTANA MAID*, assigned to the 41st Bomb Group, carried a Western Flare. *Smith*

*Miss Charlene* was the name given to the art on this B-25 Belonging to the 41st Bomb Group. *Smith*

*"MITCH" THE WITCH*, a B-25 assigned to the 41st Bomb Group, sports a young lovely witch riding on a twin tail broom. This aircraft boasts of many bombing missions and two Japanese aircraft downed. *Smith*

*NAUGHTY MARIE*, a B-25, flew with the 41st Bomb Group, Seventh Air Force. *Smith*

The 437th Bomb Squadron, 319th Bomb Group's B-25J-11, *The MILK RUN* featured the Borden Milk Company's well known logo, *Elsie the Cow*. The art was painted by David Reid who had been an employee of Borden before the war where he designed this now prominent art for Borden. *Oyster*

The 341st Bomb Group B-25, *"TEXAS GAL,"* is well armed with a 75mm cannon and four 50 cal. machine guns in its nose. *Ethell*

*"The Macon Jackpot"* was a 341st Bomb Group B-25. *Hernandez*

*Apache Princess*, a B-25, was assigned to the 501st Bomb Squadron, 345th Bomb Group. The crew of this aircraft was captured after it was shot down on 27 May 1945 in the area of Mizukami, Formosa. They spent the remainder of the war as POWs. *Watts*

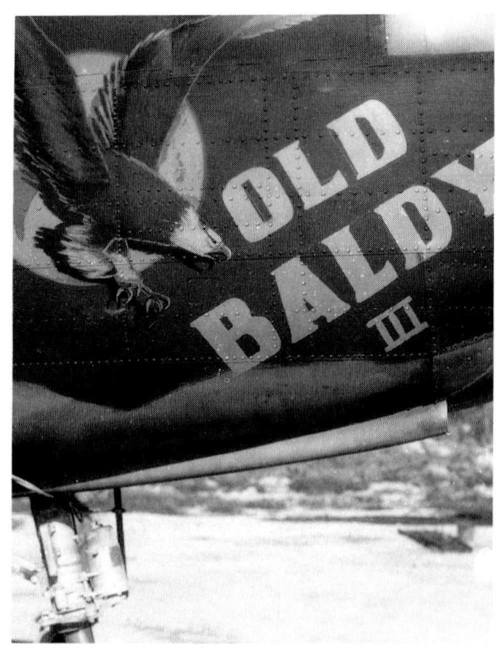


*OLD BALDY III*, a B-25J-6, serial 43-28104, belonging to the 345th Bomb Group, was in reserve at Clark Field, Philippine Islands at the time of this photo. *Smith*

A full view of a B-25 belonging to the 499th Bomb Squadron, 345th Bomb Group, gives the feeling of fierceness portrayed by the bat design on the 499th's aircraft. *Hernandez*

This front view of a B-25 belonging to the 499th Bomb Squadron, 345th Bomb Group provides a close look at the bat design carried by the aircraft of that squadron. This photo was taken on Ie Shima in August 1945. *Smith*

This early Air Apache gun ship, *SURE SHOT*, belonged to the 501st Bomb Squadron, 345th Bomb Group. *Smith*

This B-25 belonging to the 345th Bomb Group became the first American aircraft to land on Philippine soil 28 October 1944. *LAZY DAISY MAE*, a B-25J-11, serial 43-36012, disappeared over Sibuyan Island 9 January 1945 after a mission support over Lingayen Landing in Luzon.

With an Indian maid for the art on the B-25J-11, serial 43-36172, the name *WHITE WING* completes the Native American motif on this 345th Bomb Group aircraft. *Smith*

The art named *"Cherokee Princess"* was applied to a B-25H belonging to the 498th Bomb Squadron, 345th Bomb Group. This aircraft carried the squadron marking of the falcon head on the nose of the aircraft. A portion of this insignia is apparent in this photo. The *"Cherokee Princess"* was painted by Sgt. Charles Pushetongqua, an American Indian. *Smith*

In some nose art the artist revealed a religious belief. The B-25, *AVE MARIA* has a deeply religious tone. At the time of this photo the Twelfth Air Force aircraft had been on fifteen bomb missions. *Campbell*

The B-25, *BURMA VIRGIN*, shows a young oriental female riding on the back of a tiger. *Hernandez*

This art was found on a B-25 in the Pacific Theater. *MAFB*

This art was called *MMR* which stood for Mary Margaret Rustin, the girlfriend of the pilot of this B-25J-11. *Grantham*

The Twelfth Air Force B-25 called *POOPSIE* had little Poopsies over each of its sixty-six bomb mission markers. *Campbell*

*STRIPPED FOR ACTION* was on a B-25 in the Southwest Pacific. *Hernandez*

*BALLS O' FIRE*, has the often seen character, *Snuffy*, pulling Hitler behind him by a string. *Campbell*

# B-26's

The B-26, *LONESOME POLE-CAT*, flew 101 sorties for the 319th Bomb Group in the Mediterranean Theater. It was among one of the fifteen aircraft that as a group racked up a total of 1,716 sorties with a total of 12,248.5 flying hours. *Baker*

*The Repulsive Rabbit* was also one of the more famous aircraft of the 319th Bomb Group. This B-26, serial 41-31590, piloted by Capt. W.C. Wood, flew a total of 125 sorties. The bombardier was George "Squirrel" Cunningham. This photo was taken at Base Air Depot #2, Watton, England. *Baker*

This photo of *El Diablo*, a B-26 assigned to the 319th Bomb Group, was taken at Base Air Depot #2. *Lobotzke*

*SAN ANTONIO ROSE/It's a Lemon*, was a B-26 assigned to the 391st Bomb Group. The lemon on the side says it all about how the crew felt about their aircraft. *Hernandez*

*gayle*, a B-26, had its own little gremlin who is shown here boring a hole into the aircraft. *gayle* was piloted by 1st Lt. R.C. "Poppa" Foote of the 391st Bomb Group. *Hernandez*

Another of the 391st Bomb Group's Marauders carried the initials, *T.S.*. *Hernandez*

*WABASH Cannonball* was the name given this B-25 assigned to the 491st Bomb Squadron, 341st Bomb Group. The 491st operated chiefly against enemy transportation in Central Burma by bombing bridges, locomotives, and railroad yards, slowing the movement of supplies to Japanese troops. *Ethell*

This line-up of B-25s bearing the Panther insignia on their noses belonged to the 405th's sister squadron, the 822nd, also assigned to the 38th Bomb Group in the Pacific. These aircraft carried a yellow identification stripe on the bottom of the tail. *Soderlund via Copeland*

377

This Pacific B-25 carries the Green Dragon insignia seen on the aircraft of the 405th Bomb Squadron, 38th Bomb Group. The top of the tail carried a green identification stripe. Note the red of the dragon's mouth extends to and includes the machine guns which appear to be part of the dragon's tongue. *Soderlund via Copeland*

This o/d B-25 is wearing the insignia of the 491st Bomb Squadron, 341st Bomb Group. The 341st Bomb Group joined the combat in China on 10 January 1943 and was there until 25 July 1945 after the end of hostilities. The group received a DUC for developing and using the technique of bombing a target from a gliding airplane (glip bombing). *Ethell*

This is the same insignia of the 491st Bomb Squadron, but this time appears on a natural metal B-25. *Ethell*

*WHODUNIT?* The 2nd, a B-25, sports a somewhat pregnant female which explains the name given to the aircraft. This photo was taken in New Guinea.
*Soderlund, via Copeland*

*Dream Girl*, serial 44-34423, was another of the B-26Cs assigned to the 34th Bomb Squadron, 17th Bomb Wing. *Mikesh*

*Sweet Miss Lillian* was a B-26 Invader, serial 44-34334, assigned to the 37th Bomb Squadron, 17th Bomb Wing. This photo was taken at Pusan, Korea in 1952 during the Korean Conflict. The B-26, a medium bomber, had its first taste of combat during World War II. It was, however, called upon once again when war broke out between the Communist North Koreans and South Korea. *Mikesh*

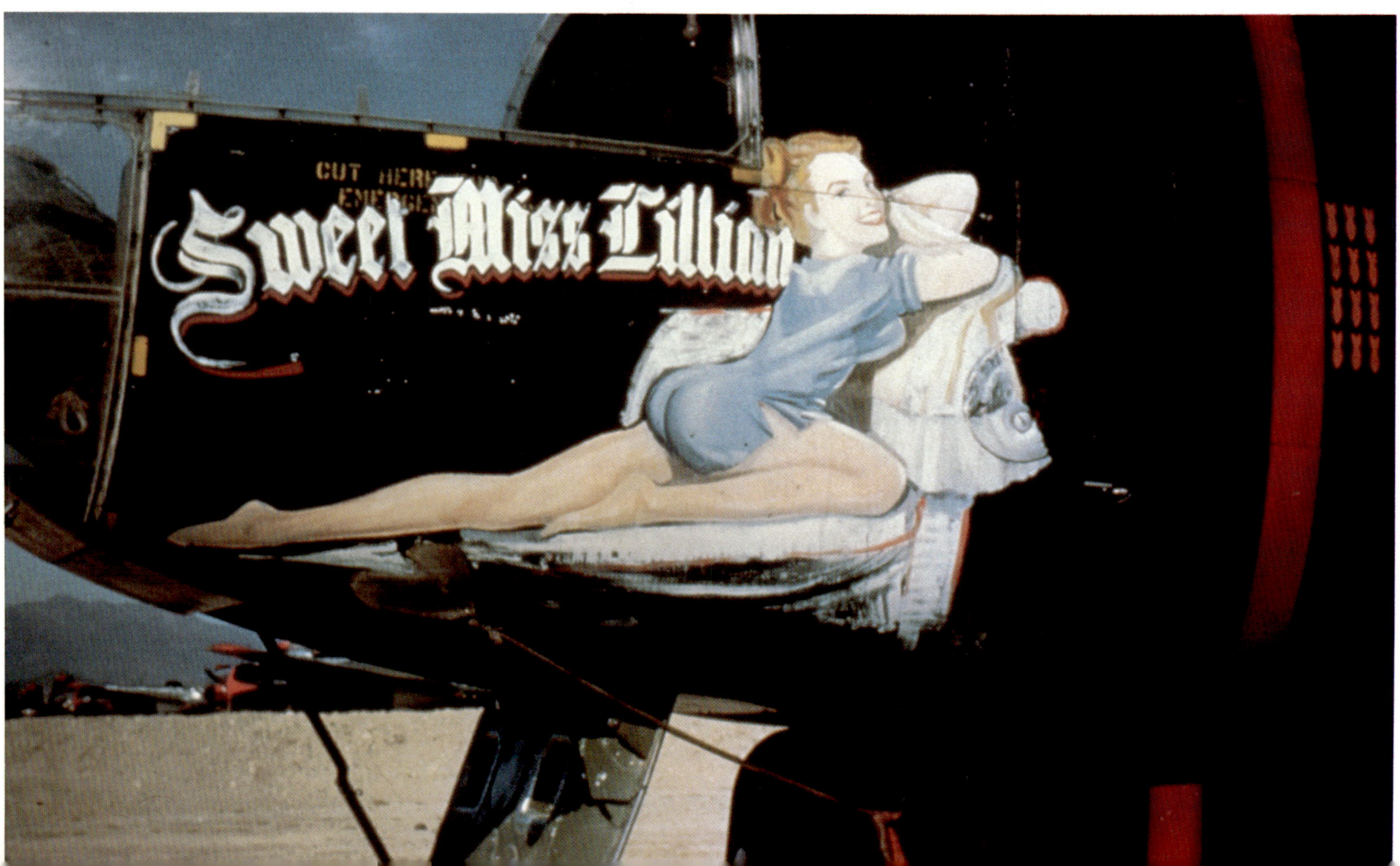

*Monie*, a B-26C, serial 44-34517, was the aircraft of Capt. Robert Mikesh who flew for the 37th Bomb Squadron, 17th Bomb Wing. Pusan, Korea, 1952. *Mikesh*

*Black Widow*, a B-26C, also served with the 37th Bomb Squadron. The 37th, who cut its teeth in both the ETO and MTO during World War II, received the Korean Presidential Unit Citation during its participation in the Korean Conflict. *Mikesh*

*No Sweat*, a B-26C, serial 44-34344, is seen here in a line of aircraft belonging to the 728th Bomb Squadron, 452nd Bomb Wing, Pusan, Korea 1951/52. *Mikesh*

G
434344
BC-34

*MY DRAGON ASS*, a natural metal B-25, served in the China-Burma-India Theater. Note the Asian flare given to the lettering on the name. *Horton*

*BIG TIME OPERATOR*, a B-25, served in the CBI. At the time of this photo the aircraft had completed twenty-six missions during which they shot down three Japanese aircraft. *Horton*

*"Jersey Bounce"* was the name given this Korean nose art found on a 452nd Bomb Wing B-26. *Campbell*

*SWEET SUE* was a 37th Bomb Squadron B-26. *Campbell*

The Douglas B-26B, *BROWN NOSE*, assigned to the 730th Bomb Squadron, 452nd Bomb Wing, had eight fixed .50 caliber machine guns with which it strafed ground targets. Note also the many bombing missions completed by this aircraft. The mission markers extend to the nose gear doors. *Mikesh*

NOSE

*"MY DRAGON ASS"* more than likely explained the way the pilot and crew felt about the number of hours spent flying the hump in the China-Burma-India Theatre. This C-47 was assigned to the 11th Combat Cargo Squadron, 3rd Combat Cargo Group which was charged with the delivery of troops and supplies to men of the CBI. *Bielauskas via Ethell*

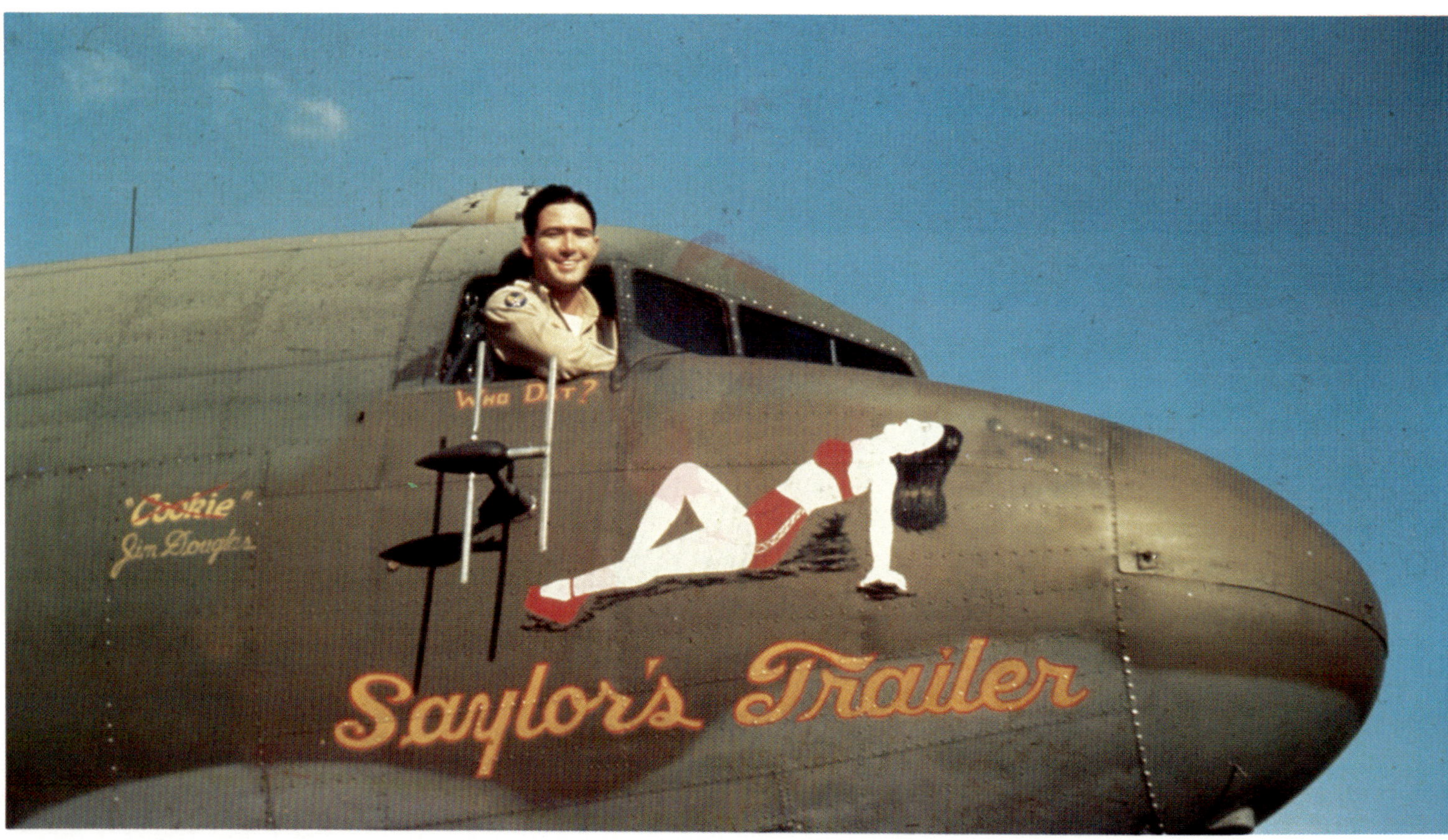


The smile of a happy Jim "Cookie" Douglas, is seen here in the cockpit of *Saylor's Trailer*, a C-47, also assigned to the 3rd Combat Cargo Group in Burma. This art was inspired by the name of the ships pilot, George Saylor. *Bielauskas via Ethell*

*PUSSY Foot Express* is seen here in the common olive drab paint scheme which was so indigenous to the CBI. This C-47 was assigned to the 3rd Combat Cargo Group. *Bielauskas via Ethell*

*MOANIN MONA* was a C-47 belonging to the 3rd CCG, China-Burma-India Theatre. We can see here the washed out faded olive over gray paint scheme weather by monsoons and countless hours in the tropical sun. *Bielauskas via Ethell*

This AC-47, more the exception to the rule in this book, is seen in a newer darker type of camouflage. Assigned to the 4th Aircraft Control Squadron in Pleiku, S. Vietnam, it served as a testimonial to the rugged abilities of the "Goonie Bird." Names like *Spooky* became an enigma in the Vietnam War, as such terms were bought to reality along with slogans such as, *Death from Above* and *Puff the Magic Dragon*. The C-47 saw more than 50 years of military service and many not so war weary C-47s still fly today, a tribute to Douglas Aircraft Company. *Mikesh*

The well placed name on the B-26, *Garden State Express*, along with the state of New Jersey served as censoring for the art on the nose of this aircraft. *Campbell*

*FAT CAT*, a C-54 cargo plane and obviously the personal transport of a Brig. General is seen at K-14 Korea in 1953. The Pilot was Captain B.W. Luster and his Crew Chief was T/Sgt. E.I. Kirtpatrick. *Jones*

*MAD BULL*, a B-26, was assigned to the 391st Bomb Group. The art was an angry bull preparing to charge. *Hernandez*

*LITTLE PINK PANTIES*, a B-26, serial 42-107841, was assigned to the 391st Bomb Group. Their mission markers show numerous bomb missions flown by this Marauder. *Hernandez*

*POP'S WAGON*, a B-26, was piloted by Lt. William "Big Willie" Hollis. Hollis flew with the 391st Bomb Group. *Hernandez*

*"VAGABOND,"* a B-26, assigned to the 391st Bomb Group, shows a hobo on the road to Berlin, 600 miles away according to the road sign. In the opposite direction was the U.S.A. at an unidentified distance. At the time of this photo this aircraft had flown fifty-seven missions. Note the star within the mission markers. *Hernandez*

*"Silver Dollar,"* a B-26, was piloted by Lt. A.P. Schiro. A silver dollar with wings adorned this 391st Bomb Group Marauder. *Hernandez*

*Baby Doll III*, a B-26, was piloted by 1st Lt. R.J. Snyder of the 391st Bomb Group. This small tyke rides a bomb titled, *Berlin Express*, as if it were a toy horse. *Hernandez*

This natural metal B-26 was named, *Dode Lee's Memphis Blues*. It was assigned to the 391st Bomb Group. *Hernandez*

Sitting in an island paradise under a palm tree is the subject of this art, *DON The BEACHCOMBER*. This is a B-26 assigned to the 391st Bomb Group. *Hernandez*

The 391st Bomb Group's B-26, *ZOMBIE*, carried this gruesome creature. At the time of this photo *ZOMBIE* had been on sixty bomb missions. *Hernandez*

*Pink's Lady II*, a B-26, assigned to the 391st Bomb Group, sports a glass of bubbly. *Hernandez*

*SKYHAG*, a B-26, was piloted by Lt. F.J. Majka. The art on this 391st Bomb Group aircraft is of a witch dropping a bomb. Note the swastikas on each of the bombs marking bomb missions against the Fuhrer. *Hernandez*

A wolf representing each member of the crew is seen on the B-26, *The WOLF PACK*, assigned to the 391st Bomb Group. *Hernandez*

*The Huntress*, a B-26, serial 42-95832, was piloted by Capt. Bush of the 391st Bomb Group. *Hernandez*

*"LILLY COMMANDO,"* a B-26, was assigned to the 391st Bomb Group. This photo was taken August 1944 while the group was stationed at Roye/Amy, France. *Hernandez*

*HITCH HIKER*, a B-26B-50-MA, serial 42-95842, was assigned to the 391st Bomb Group. It had flown twenty-two missions at the time this photo was taken. *Hernandez*

*Countess II*, a B-26, was assigned to the 391st Bomb Group. The art on this aircraft was small enough to be painted on one section of the metal. *Hernandez*

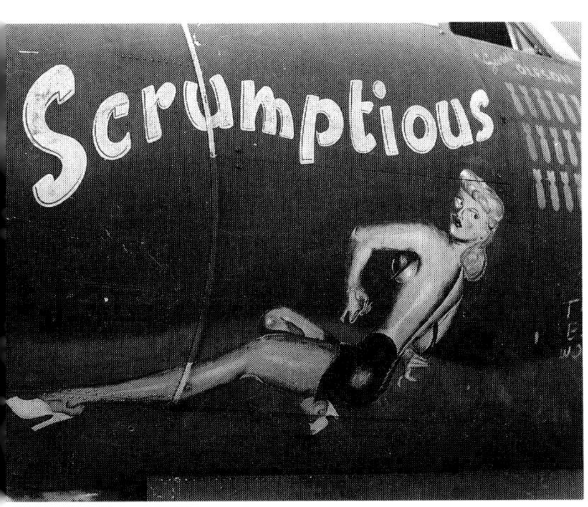


*Scrumptious*, a B-26, assigned to the 391st Bomb Group, carried an often seen pinup. *Hernandez*

*Sleepy-time gal*, a B-26, was piloted by 1st Lt. C.G. Aldous of the 391st Bomb Group. *Hernandez*

*"LADY GODIVA"*, a B-26-C-45-MO, serial 42-107811, was assigned to the 391st Bomb Group. *Hernandez*

*B-BABY*, a B-26B-50-MA, serial 42-95950, was assigned to the 391st Bomb Group. *Hernandez*

*Skeeter*, a B-26, was piloted by 1st Lt. R.B Dearling of the 391st Bomb Group. *Hernandez*

*POWERFUL KATRINKA*, a B-26, was piloted by Lt. "Geep" Garwick of the 391st Bomb Group. *Hernandez*

*"SWEE'PEA"* was a 391st Bomb Group B-26. The Popeye Cartoon character is seen here floating on a cloud aiming a machine gun. *Hernandez*

This art was on a B-26 assigned to the 391st Bomb Group. *Hernandez*

*Johnny's Jokers*, a B-26G-25-MA, serial 44-68122, was piloted by Lt. Jack Avooski. The aircraft carried the name of the CO, Maj. J.C. Ruse, which is possibly who the aircraft was named for. *Ethell*

*Pappy's PRAM*, a B-26, serial 41-31002, shows an old-timer pushing a baby buggy containing a load of bombs. Ethell

Pictured with *GREMLIN HQ*, is an entire group of Royal Air Force personnel posing with a captured Italian "Savoia-Marchetti" bomber. The nickname for this bomber is probably derived from the mechanical faults known to have plagued this particular aircraft. *Stacy*

A sad little penguin adorned the B-26, *BUTCH*, assigned to the 391st Bomb Group. *Hernandez*

*WINE WOMEN & So Long*, this B-35 Hudson, in South America during the later part of World War II, bears the insignia of the 4th Combat Mapping Squadron. The mapping squadrons were largely responsible for the photo recon role as well. *Shaw*

*"WE'LL SLAP THE JAP OFF THE MAP"*, stares at the viewer with the true feelings towards the enemy back in 1944 when this photograph was taken. Seen here on the after fuselage of a B-35 Hudson is yet another place for art work to appear. This aircraft was assigned to the 4th Combat Mapping Squadron. It was the common practice in the U.S. Navy for the art work to be applied in this same location on their operating "Venturas." The Ventura and Hudson were the same aircraft with only minor differences. *Shaw*

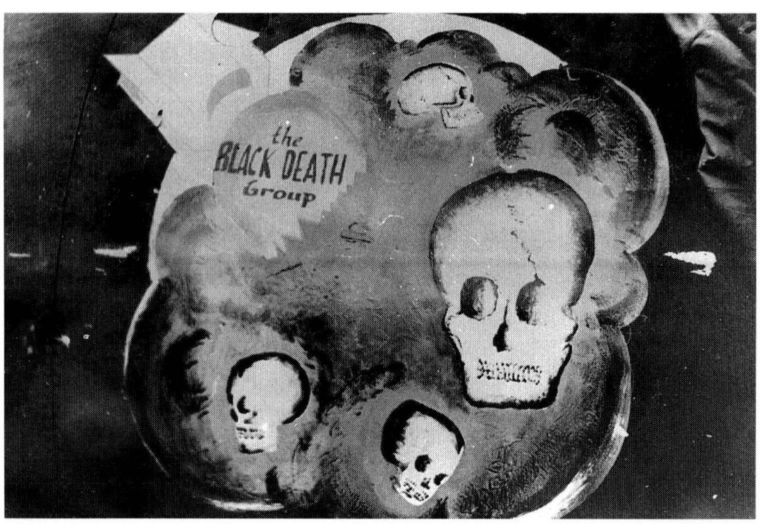


This cloud of smoke containing skulls was seen on the B-26 titled, *the BLACK DEATH Group*. *Campbell*

This photo of the B-26, *WOLF PACK*, was taken at Bradley Field, Connecticut. The photo was shot by Col. Mike Moffitt. *Moffitt*

*IT'S A "TOUGH RACKET,"* a PV-1 attached to VB-144 is seen after a wheels up landing on Tarawa in 1944 – also notice the light flak damage on the right hand side of the name. *Ethell*

This nameless PV-1 Ventura, #730, was assigned to VB-144 on Tarawa during 1944. *Ethell*

*JIMMY JUNIOR*, a PV-1, #34877 was assigned to VB-144. This Ventura was stationed on Tarawa in 1944 in the capacity of an anti-submarine warfare deterrent. *Ethell*

*Left: THE TURTLE* seen after the historic flight in which this P2V flew a record breaking 11,237 mile flight from Perth, Australia, to Columbus, Ohio. The Neptune aircraft was piloted and commanded by Commander T.D. Davies. The co-pilot was Commander E.P. Rankin. The record was previously set by the B-29 Superfortress named *Pacusan Dreamboat*. The P2V Neptune was later referred to as *The Truculent Turtle* and found her final resting place on permanent display at the Naval Aviation Museum at Pensacola, Florida where she sits today. *Martin*

*Below left: THE THING*, a P2V Neptune, was assigned to Fast Attack Squadron 112. *The Thing* had the distinction of performing the 15,000th landing for FASRON 112's Transitional Training Department. The pilot Lt. L.A. Moran was congratulated with the rest of his crew on the record breaking landing. *The Thing* was the second oldest operating P2V in the Navy and had an accident free record. The name was taken from the movie of the same name which starred James Arness in the starring role. ADE1 Philip E. Rich designed and painted *THE THING*. *Rich*

*Below: 'TAINT WHAT CHA DO, HITS THE WAY YA DO IT* accompanied the art on this B-35 Hudson stationed in South America 1944. *Shaw*

Miss Beryl

# Chapter Two

# Heavy

When the word Liberator is heard one thinks of liberation and freedom. The Consolidated B-24 certainly lived up to this name as it fought for freedom and liberty in all theaters of World War II. This rugged aircraft proved her worth in the frigid wind swept skies over Germany and France, the hot arid air of the desert where it was constantly pelted by blowing sand, and the wet, sticky, tropical climate of the Pacific. However, the B-24 is largely remembered and praised for its costly missions flown against the life flow of the German war machine – the oil refineries of Ploesti at a cost of 339 bombers lost along with their crews.

The B-24 was more widely produced than any of its contemporaries with over 18,000 built. Its range to target was longer than any other U.S. heavy bomber until the advent of the B-29. The Army Air Corp flew the B-24 under the name Liberator, while the Navy flew the PB4Y-1, also called Liberator, and its single tail sister the PB4-2, dubbed the Privateer. The British, flew the B-24 under the designation LB-30.

The B-24, in fact, operated in more than just a bomber role. It flew in missions nicknamed "The Carpetbaggers", during which agents and supplies were dropped behind enemy lines. It was also designated C-87 and served as a cargo vessel transporting fuel, supplies, and troops to every conceivable location.

The B-24 stands as a tribute to the thousands of men and women who built, modified or otherwise supported this wonderful legend in aviation history.

*Miss Beryl*, a B-24L, serial 44-40832, was assigned to the 308th Bomb Group. Along side the multiple bomb mission markers, carried by *Miss Beryl*, are eight freighters. One is followed by a question mark denoting one possibly got away. *Beitling*

*Ruff Knights*, a B-24J, serial 44-40550, was assigned to the 42nd Bomb Squadron, 11th Bomb Group after its transfer from the 38th Bomb Squadron, 30th Bomb Group. To get the full effect of the art you need to examine both sides of the aircraft. As can be seen it is a continuing mural of cartoon characters running around the aircraft. *Ruff Knights* flew 35 missions. *Crow*

# B-24's

*TRICKY MICKY*, a B-24M, serial 44-50960, was a member of the 98th Bomb Squadron, 11th Bomb Group. The 98th's first assignment was Hickam Field from which it flew search missions. On the opposite side of the plane was the 98th's insignia, a diving bomb. *Crow*

*I'SE A ROYAL HAWAIIAN*, a B-24J, serial 42-41198, was assigned to the 26th Bomb Squadron, 11th Bomb Group. *Cleveland*

*KANSAS CYCLONE*, B-24J, serial 42-73025, was a member of the 26th Bomb Squadron, 11th Bomb Group. The art portrays a Japanese interrupted by an act of nature common in Kansas – a cyclone. *Cleveland*

*PATCHES*, a B-24, serial 42-73018, was assigned the 431st Bomb Squadron, 11th Bomb Group. This very busy picture portrays the many functions of the crew. *Patches* had been known as *"The Captain and The Kids."* After repairs were made the name changed. *Cleveland*

*BARREL HOUSE BESSIE*, a B-24D, serial 42-73027, was assigned to the 98th Bomb Squadron, 11th Bomb Group. This is not the original painting. *Barrel House Bessie* was repainted after repairs were made. *Cleveland*

*Angel Face the 2nd*, a B-24, serial 44-50217, was assigned to the 98th Bomb Squadron, 11th Bomb Group. *Angel Face the 2nd* was retired after 101 missions in June of 1944. *Cleveland*

*LITTLE DUMBO*, a B-24J, serial 44-40143, served both with the 26th and 431st Bomb Squadrons, 11th Bomb Group. The Disney character in varying poses was seen on several of the planes belonging to the 11th Bomb Group and many others. *Cleveland*

*Millie's daughter*, a B-24J, serial 42-109876, was assigned to the 11th Bomb Group. *Cleveland*

*Naughty NANETTE*, a B-24J, serial 44-50961, was assigned to the 42nd Bomb Squadron, 11th Bomb group. *Cleveland*

*ROYAL FLUSH*, a B-24J, was assigned to the 42nd Bomb Squadron, 11th Bomb Group. *Cleveland*

*READY TEDDY II*, a B-24, serial 42-73008, was assigned to the 98th Bomb Squadron, 11th Bomb group. *Cleveland*

*RITA*, a B-24, was assigned to the 42nd Bomb squadron, 11th Bomb Group. *Cleveland*

*SUPER BABY*, a B-24, serial 42-109873, was assigned to the 98th Bomb Squadron, 11th Bomb Group. *Cleveland*

*SUPER MAN*, a B-24D, serial 41-11938, was assigned to the 26th Bomb Squadron, 11th Bomb Group. *Cleveland*

*"LIL' AUDREY" GROWS UP*, a B-24, serial 42-73016, was assigned to the 431st Bomb Squadron, 11th Bomb Group. It was the first aircraft in the 7th Air Force to complete 100 missions. *Inset:* This side of *Lil' Audrey* lists the statistics on the 100 missions flown by the aircraft. Pictured is a B-24 flying over Tokyo dropping fifteen bombs carrying the names of the fifteen targets to Tokyo. *Cleveland*

*LADY FROM HADES*, a B-24, serial 44-40302, was assigned to the 26th Bomb Squadron, 11th Bomb Group. *Cleveland*

*PACIFIC VAMP, (Mrs. Pacific Tramp)*, a B-24, serial 41-24168, was assigned to the 98th Bomb Squadron, 11th Bomb Group. It had earlier carried the name *Valhalla* and served with the 26th Bomb Squadron. *Cleveland*

*GUN SITE*, a B-24J, serial 42-73005, was assigned to the 26th Bomb Squadron, 11th Bomb Group. *Cleveland*

*Sky's Delight*, a B-24J, serial 44-40673, was assigned to the 431st Bomb Squadron, 11th Bomb Group. *Cleveland*

*COCONUT QUEEN*, a B-24, serial 42-72992, was assigned to the 98th Bomb Squadron, 11th Bomb Group. *Cleveland*

*K Lucy II*, a B-24J, serial 42-109869, was assigned to the 26th Bomb Squadron, 11th Bomb Group. *Cleveland*

*The UNINVITED*, a B-24, serial 44-40360, was assigned to the 98th Bomb Squadron, 11th Bomb Group. Note the star-within-a-star on the nose wheel cover. *Cleveland*

*Wabbit Twansit*, a B-24J, serial 42-73006, was assigned to the 431st Bomb Squadron, 11th Bomb Group. Bugs Bunny and Porky Pig, Warner Brothers' characters, are riding a rocket for transportation. *Cleveland*

*CHUTE THE WORKS*, a B-24J, serial 44-40302, was assigned to the 431st Bomb Squadron, 11th Bomb Group. *Cleveland*

*DRAGON LADY*, a B-24J, serial 44-40670, was assigned to both the 98th Bomb Squadron and the 42nd Bomb Squadron, 11th Bomb Group. *Crow*

*TROPIC KNIGHT II*, a B-24M, was assigned to the 11th Bomb Group. *Crow*

*WISHFUL THINKIN'*, a B-24L, was assigned to the 11th Bomb Group. *Crow*

*HEAVEN CAN WAIT*, a B-24L, serial 44-73496, was assigned to the 98th Bomb Squadron, 11th Bomb Group. *Crow*

*MISS SHERRY*, a B-24L, serial 44-40710, was assigned to the 42nd Bomb Squadron, 11th Bomb Group. *Crow*

*NANCY*, a B-24L, serial 44-41498, was assigned to the 42nd Bomb Squadron, 11th Bomb Group. *Crow*

*MINNIEHAHA*, a B-24L, was assigned to the 98th Bomb Squadron, 11th Bomb Group. *Crow*

*Rough Knight*, a B-24M, was assigned to the 11th Bomb Group. *Crow*

*HAP-N-HANK*, a B-24L, serial 44-50145, was assigned to the 11th Bomb Group. This aircraft was the 5,000th B-24 built by Ford. The *Hap* was for General "Hap" Arnold while the *Hank* was for Henry Ford. *Hap-N-Hank* completed eight missions before the end of the war and was returned to the states. *Crow*

*UNDER EXPOSED*, an F-7, serial 42-73056, assigned to the 20th Combat Mapping Squadron, sports one of the classic works of Merkling. *Malcom*

*PATCHED UP PIECE* was a name that described this F-7, of the 20th Combat Mapping Squadron after a few missions. This same art was seen on the aircraft after the paint was stripped, and it was a bare metal bird. *Malcom*

*Curly Bird*, a B-24J, serial 44-40683, was assigned to the 819th Bomb Squadron, 30th Bomb Group. *Crow*

*Ole' TOMATO*, an o/d B-24, was assigned to the 22nd Bomb Group. *Malcom*

*Redhot Ridinhood*, a B-24, serial 44-4985, was assigned to the 33rd Bomb Squadron, 22nd Bomb Group, Fifth Air Force. This aircraft crashed 28 February 1944. *Malcom*

PROJECT NO. 98195
U.S. ARMY B-24-L-11 F.O.
AIR FORCE SERIAL NO. 44-49528
FIRE EXTI
BEA

*BEAUFORT BELLE*, a B-24L, serial 44-49528, was assigned to the 55th Weather Squadron. *Crow*

*THAR SHE BLOWS III*, a B-24, serial 44-41468, was assigned to the 30th Bomb Group. She was returned to the states upon retirement. *Crow*

*WILD ASS RIDE*, a B-24, serial 44-41615, was assigned to the 38th Bomb Squadron, 30th Bomb Group. It was later transferred to the 42nd Bomb Squadron, 11th Bomb Group. *Crow*

*The CAPTAIN and The KIDS*, a B-24, serial 44-40518, was assigned to both the 392nd and 819th Bomb Squadrons, 30th Bomb Group. *Crow*

*RAT POISON*, a B-24, serial 41-29560, served with the 34th Bomb Group and was later transferred to the 392nd Bomb Group. It was salvaged 16 March 1945. This photo was taken at Base Air Depot #2, England. *RAT POISON* was an early 24 with the o/d paint and only showed three bomb missions at the time of this photo. *Lobotzke*

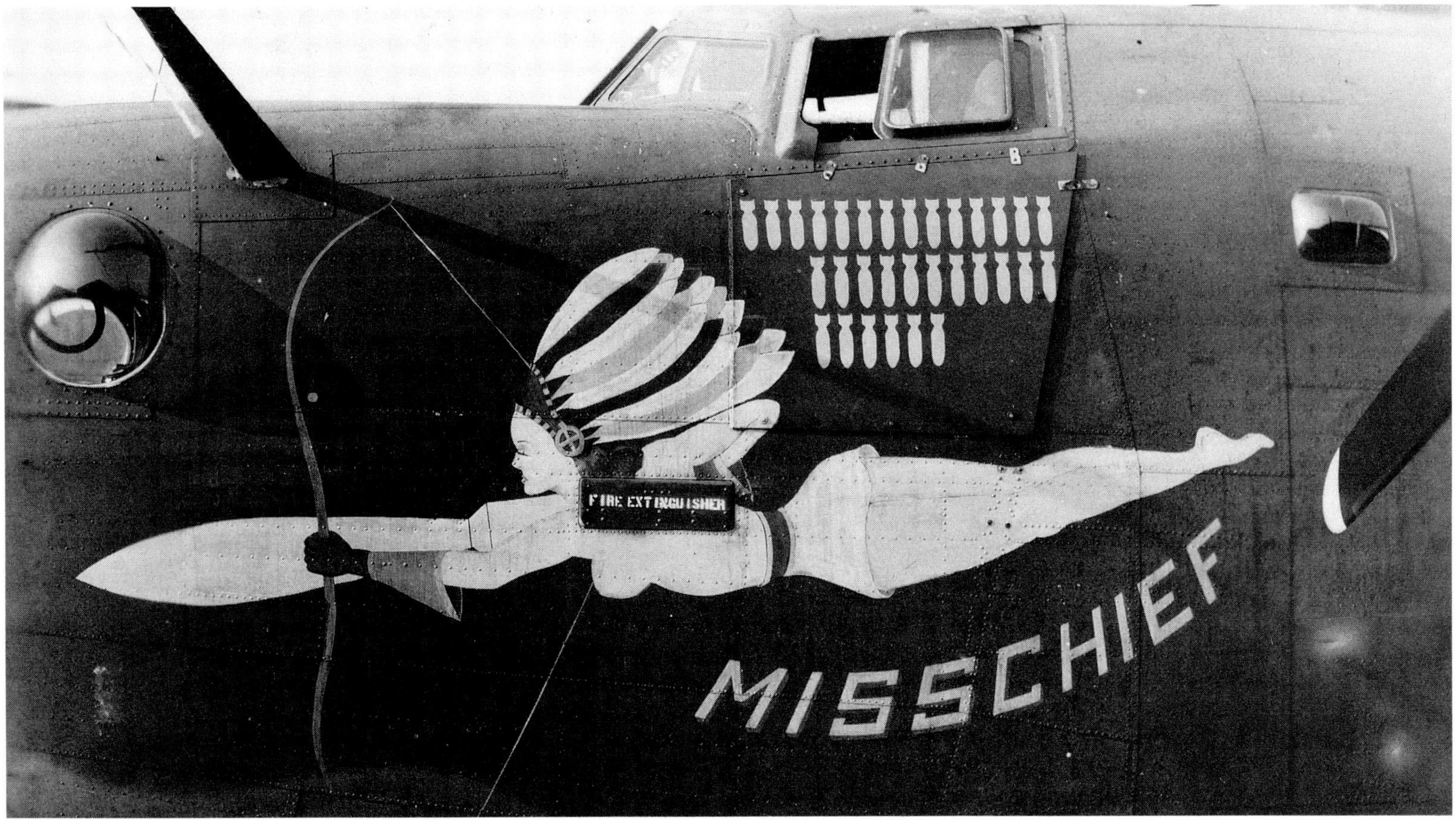


*MISSCHIEF*, a B-24, serial 42-94880, was assigned to the 34th Bomb Group. This photo was taken at Base Air Depot #2 during armor modifications. *Lobotzke*

The B-24, serial 44-50768, *"MILLION $ BABY"*, served with 403rd Bomb Squadron, 43rd Bomb Group. The 43rd served in the Southwest Pacific with the Fifth Air Force. The group experimented with skip bombing and used this technique for attacks on Japanese vessels during the Battle of the Bismarck Sea. During that battle, with repeated air attacks they destroyed a large enemy convoy carrying reinforcements to New Guinea. For this action they received a Distinguished Unit Citation. *Planes of Fame Museum*

*LUCKY LUCILLE*, a B-24M, serial 42-41224, was assigned to the 65th Bomb Squadron, 43rd Bomb Group. *Malcom*

One of the most famous paintings of the 43rd Bomb Group was *COCKTAIL HOUR*. This B-24L, serial 44-40428, flew with the 64th Bomb Squadron, 43rd Bomb Group. *Planes of Fame Museum*

The B-24, serial 44-40973, named *THE DRAGON AND HIS TAIL*, is another very famous aircraft belonging to the 64th Bomb Squadron, 43rd Bomb Group. *Planes of Fame Museum*

*DOUBLE TROUBLE*, a B-24L, serial 42-40358, was assigned to the 320th Bomb Squadron, 90th Bomb Group. *Crow*

*FLAK ALLEY*, a B-24D, serial 41-42225, flew the Ploesti missions with the 68th Bomb Squadron, 44th Bomb Group. The name *Flak Alley* had special meaning for those men and aircraft who flew in low over the highly protected oil refineries of Ploesti. Note the horizontal bomb mission marker denoting low-level bombing. Before 24 February 1944 when this aircraft was shot down by fighters near Gotha, it racked up a total of 36 mission markers and seven Nazi flags. The two ducks denote decoy missions. *Beitling*

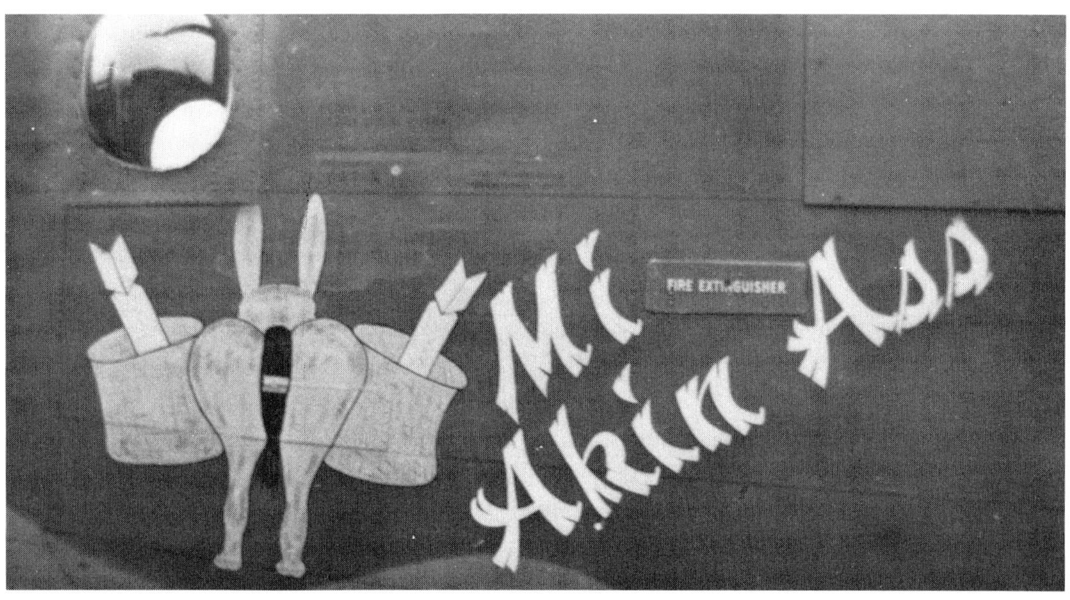


*Mi Akin Ass*, a B-24J, serial 42-94846, was assigned to the 67th Bomb Squadron, 44th Bomb Group. After completing its combat tour, *Mi Akin Ass* was retired to the states. *Ethell*

With a bomb for a throne, this majestic blonde, wears on the neck of her robe and the tip of her scepter, a unique form of the skull and cross-bones belonging to the 90th Bomb Group. The 90th substituted the cross-bones with cross-bombs. The *QUEEN MAE* was an early B-24J with the narrow navigator windows. This aircraft flew with the 319th Bomb Squadron. *Campbell*

This American lass, clad only in shorts, swings her scimitar at the fleeing leaders of the Axis, Hitler and Tojo. Note the Jolly Rogers insignia on her shorts. *THE BUTCHER'S DAUGHTER*, a B-24J, serial 44-40190, has a impressive score board with seven Jap flags leading its numerous bomb mission markers. She flew with the 319th Bomb Squadron, 90th Bomb Group. *Campbell*

This very popular Varga Girl decorated the B-24, serial 44-40185, *QUEEN OF HEARTS*. Adding to its appearance is a smiling shark mouth. The *QUEEN OF HEARTS* flew with the 320th Bomb Squadron, 90th Bomb Group. Note the valentine shaped hearts on the trailing vail which made this Varga Girl unique to this aircraft. *Campbell*

This 90th Bomb Group B-24J, serial 42-109987, named *PRETTY BABY*, sports a lovely lady floating on a cloud. Behind her are the bomb mission markings. *Campbell*

*PAPPY'S PASSION*, a B-24J, serial 42-100222, was assigned to the 319th Bomb Squadron, 90th Bomb Group. *Campbell*

*Shamrock Sherry*, a B-24J, flew with the 319th Bomb Squadron, 90th Bomb Group. She carries a four leaf clover for luck. *Campbell*

The B-24M, serial 44-40229, *FORM 1-A/HARRY S. TRUMAN*, flew 150 missions with the 321st Bomb Squadron, 90th Bomb Group. *Campbell*

*V. . . -SURE POP*, a B-24J, serial 42-41073 was assigned to the 90th Bomb Group. *Malcom*

*Miss Kiwanis*, featuring Hollywood's own Barbara Stanwyck, was one of the B-24s assigned to the 319th Bomb Squadron, 90th Bomb Group. Note the Jolly Rogers insignia on the ammo box. This is the art of Jack Clipper. *Campbell*

This aircraft appears to be in an early state of salvage. Note the missing plexiglass and loose framework in the cockpit area. *NAVEL MANEUVERS* was on loan to the 90th Bomb Group, as many different aircraft from sister squadrons and groups were due to attrition and the need to fill bomber formations to their operating capacities. *Campbell*

This was possibly the last picture taken of the 5th Bomb Group's, B-24J, *DRAGON LADY*, serial 44-49753. The war weary aircraft, realizing the end of her career, waits at Kingman, Arizona for her turn in the smelter. *Grantham*

The splendid lion's head on the nose of this B-24 was painted at Oklahoma City Air Depot at Tinker Field prior to its departure to Horsham St. Faith, England where it was assigned to the 458th Bomb Group. *Tinker AFB*

89

*OLD IRONPANTS*, a B-24, #168. *Bietling*

The letters on this shark mouth B-24, spelling *PATCHES*, were painted in different colors in keeping with the theme. There were many aircraft with this name. It would have been an appropriate name for many more aircraft who crippled in on a wing and a prayer from very costly combat missions. *Horton*

*BACHELOR'S BROTHEL*, a B-24L, serial 44-50005, was assigned to the 531st Bomb Squadron, 380th Bomb Group. *Bachelor's Brothel* set down in Japan on 2 September 1945, making it the first of the 380th's aircraft to land on Japanese soil. *Acord*

*PUDDLE JUMPER II*, a B-24M, serial 44-41946, was a member of the 30th Bomb Group. *PUDDLE JUMPER II* is credited with sinking a 5000 ton enemy freighter. It was later attacked by Japanese fighters and crashed into the sea 26 December 1944. *Acord*

*ATOMIC BLONDE*, a B-24M-25-FO, serial 44-51414, was a member of the 531st Bomb Squadron, 380th Bomb Group. *Acord*

*SLIP STREAM* was among the B-24s which had been sent to Okinawa to await final disposition. Like many of the names given to the aircraft, *Slip Stream* has a double meaning. The pilot took an aviation term and gave it a new twist. In this painting the slip stream affords the "girl watcher" a better view. *Acord*

*LIBERTY Belle*, a B-24M-15-FO, serial 44-50894, was assigned to the 529th Bomb Squadron, 380th Bomb Group. *Acord*

*'Luvablass'*, a B-24M-25-CO, serial 44-42264, spent time with both the 529th and 531st Bomb Squadrons which were with the 380th Bomb Group. *Acord*

*NIGHT and DAY* was a B-24 belonging to the 11th Bomb Group. The name was a good description of the frequency of missions flown by the aircraft of World War II. *Acord*

*The Duchess*, a B-24, serial 44-40567, flew with the 494th Bomb Group. Under the co-pilot's window is the name *Little Duke*. *Acord*

The Raven aircraft, *'Missouri Miss'*, a B-24M-1-CO, serial 44-41811, was assigned to the 530th Bomb Squadron, 380th Bomb Group. The aircraft survived the war and was sent to the smelters 21 October 1949. *Acord*

*'Lucky Strike'*, a B-24, serial 44-41876, was assigned to the 530th Bomb Squadron, 380th Bomb Group. Much like the mastheads of old protected and guided their crews, *'Lucky Strike'*, beautiful and majestic, both adorned this aircraft and protected her crew. This art is the work of James Nickloy. *Acord*

*DANGEROUS CRITTER*, a B-24, serial 44-40382, was a member of the 26th Bomb Squadron, 11th Bomb Group. *Acord*

*INNOCENCE A'BROAD*, a B-24, was assigned to the 494th Bomb Group. The inspiration for this art like so much came from calendar pin-ups. This is a sample of what was created using the art of Elvgren. *Acord*

*Liquidator*, a B-24M, serial 44-42323, was assigned to the 98th Bomb Squadron, 11th Bomb Group while on loan from the 494th Bomb Group. It was later transferred to the Far East Air Force. *Horton*

*"VIRGIN ABROAD"* a B-24M, serial 44-50941, was assigned to the 529th Bomb Squadron, 380th Bomb Group. *Acord*

*SWEET ROUTINE*, a B-24J serial 44-40527, previously known as *HOME JAMES*, was assigned to the 819th Bomb Squadron, 30th Bomb Group. This aircraft was among others which later transferred to the 11th Bomb Group. *Acord*

Gimme a Great

Using a large portion of the side of the aircraft is the art, *MABEL'S LABELS/ Gimme a great big smile!*, found on a B-24, serial 44-49853, assigned to the 64th Bomb Squadron, 43rd Bomb Group on Ie Shima. *Horton*

Don't be mislead by the name, the 819th Bomb Squadron *NIGHT MISSION*, was not a night-flier. Lt. Winton Newcomb, the pilot of this B-24M, serial 44-40532, named it in hopes of getting a break from the very heavy day missions. During its 42 missions flown with the 30th Bomb Group, that never happened. *Harding*

*Ruff Knights*, a B-24J, serial 44-40550, was assigned to both the 11th and 30th Bomb Groups during its military service. *Acord*

*TYRANNOSAURUS REX*, a B-24, serial 42-40363, sports a dinosaur, crushing a captured Japanese Zero, as it prepared to munch the pilot which hangs helplessly from its mouth. This *thunder lizard*, assigned to the 90th Bomb Group, was a real Jap killer. Note the seven Japanese flags included with the bomb mission markers. *Horton*

Appropriately named for the ten crew members, *A Wing an' 10 Prayers*, a B-24, serial 44-42378, flew with the 528th Bomb Squadron, 380th Bomb Group. The chance of surviving a tour of duty as a member of a bomb crew was poor at best. The odds were increased by the rugged aircraft being flown and their ability to return home even after sustaining severe damage. A portion of the art on this aircraft was lost when a section of skin was replaced during repairs. *Acord*

*Net Results*, a B-24M-5-CO, serial 44-41875, was assigned to the 528th Bomb Squadron, 380th Bomb Group. *Net Results* was one of the B-24s in the Pacific which was equipped with electronic counter measures and referred to as a *Raven* aircraft. Not to be confused with the F-7, they did, however, assist the combat mapping squadrons. This aircraft was sent to the RFC Smelt Depot in the U.S. on 31 October 1945. *Acord*

*YO HO HO*, and a "cage" of rum decorates *Miss Jolly Roger*, a B-24J, serial 42-41190, belonging to the 321st Bomb Squadron, 90th Bomb Group. *Campbell*

*Phyllis T. of WORCESTER* was a Pacific based B-24J which was assigned to the 319th Bomb Squadron, 90th Bomb Group. Note the three Japanese flags under the 40 bomb mission marks. This aircraft was piloted by the 319th's C/O, Capt. Charles Briggs Jr. *Smith*

*SLEEPING DYNAMITE*, a B-24J, serial 42-50839, was assigned to the 93rd Bomb Group. In the upper left hand corner you can see the armor plate installed under the co-pilot's window. There was a matching panel under the pilot's window. These provided extra protection for the men who flew the aircraft of the Eighth Air Force. *Lobotzke*

*"MAULIN' MALLARD,"* featured Donald Duck brandishing a sledge hammer on this shark mouth B-24L belonging to the 93rd Bomb Group. The bomb mission marker on the armor plate boasts of 115 sorties for this 8th Air Force aircraft. *Moffitt*

The B-24D, serial 42-40402, titled the *SANDMAN*, was a member of the 98th Bomb Group. The 98th is best know for its participation in the Ploesti Raids. In this photo Donald Duck, as the Sandman, is dropping his load of sand(?) on those below. *Hill*

*Able Mable*, a B-24J, serial 42-110081, was assigned to the 328th Bomb Squadron, 93rd Bomb Group. This young soldier is using American ingenuity to solve his problems. When faced with a lady clothed in a grass skirt he brought out the lawn mower. The picture was taken at Watton Air Depot. *Lobotzke*

This B-24M, *UBANGI BAG III*, was assigned to the 308th Bomb Group. Like many of the shark mouth aircraft, it has incorporated two pieces of art on the nose. The aircraft is named for the native, but there is also a very angry shark mouth design on the nose. *Hill*

*Esky*, a B-24M, was assigned to the 308th Bomb Group in the CBI. The Esquire character, *Esky*, is shown here in his favorite sport, girl chasing. *Campbell*

This mixture of fish, fowl, animal and female adorned the 308th Bomb Group's B-24L named, *THE SNARK*. *Campbell*

The B-24D, serial 41-24288, *Little Flower*, was assigned to the 375th Bomb Squadron, 308th Bomb Group. *Bietling*

The B-24D, serial 41-24183, *The GOON*, was assigned to the 374th Bomb Squadron, 308th Bomb Group. *Bietling*

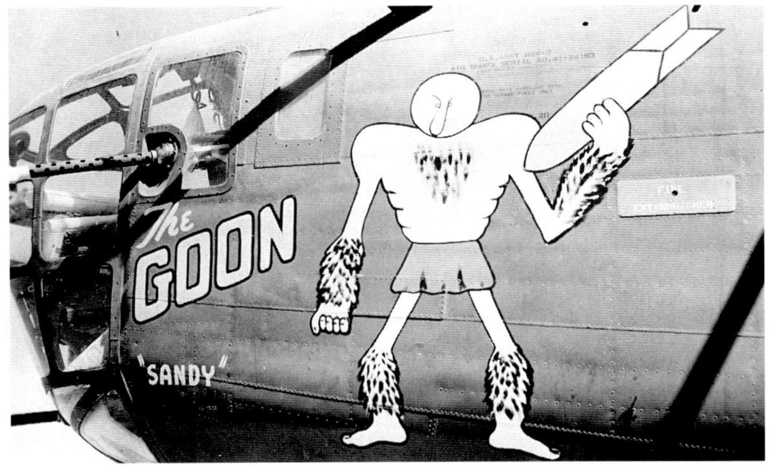


*"LAKANOOKI,"* a B-24D, serial 41-24188, was assigned to the 308th Bomb Group. Like the sailors of old, the pilots of this aircraft also had their mermaid. *Bietling*

*MISS CARRIAGE*, a B-24D, was one of the B-24s assigned to the 308th Bomb Group. *Bietling*

*AXIS NIGHTMARE*, a B-24D, serial 41-24138, was assigned to the 373rd Bomb Squadron, 308th Bomb Group. *Bietling*

*TAYLOR MAID*, a B-24J, serial 44-41294, was assigned to the 308th Bomb Group. *Hill*

*INNOCENT INFANT*, a B-24L, serial 44-49649, was assigned to the 308th bomb Group. *Hill*

*CALAMITY JANE*, a B-24M, serial 44-42019, was assigned to the 374th Bomb Squadron, 308th Bomb Group. *Hill*

Named for a popular Milton Caniff comic strip character, the B-24M, *MISS LACE*, serial 44-42133, belonged to the 308th Bomb Group. *Hill*

*LITTLE EGYPT*, a B-24L, serial 44-49569, was assigned to the 308th Bomb Group. *Hill*

*KING'S "X"* was another of the B-24s displaying a shark mouth accompanied by another predominant piece of art. This 308th Bomb Group aircraft, serial 44-40584, flew with the 375th Bomb Squadron. *Hill*

*"SETTIN PRETTY,"* at the time of this photo, had flown the hump thirteen times. This B-24L, serial 44-41429, was assigned to the 373rd Bomb Squadron, 308th Bomb Group. *Hill*

*"WILLIE MAK'ER"* was a B-24J, serial 44-40271, assigned to the 308th Bomb Group. *Hill*

*"OLD RELIABLE"* was another of the 308th Bomb Group's B-24s. It was serial 44-41437. Hanging on the outside of the outhouse was a Sear's catalog. *Hill*

This lovely lady adorned the B-24J, serial 42-100269, *"AVAILABLE,"* belonging to the 308th Bomb Group. *Hill*

*AMERICAN BEAUTY*, a B-24J, serial 44-41251, was assigned to the 375th Bomb Squadron, 308th Bomb Group. *Hill*

*GEORGIA PEACH*, a B-24, serial 42-7445, was among the B-24s belonging to the 308th Bomb Group. *Hill*

This hula dancer was on the B-24, *Hilo Hattie*, belonging to the 308th Bomb group. *Hill*

Sporting a dancing señorita is the B-24J, serial 42-73305, *Rose of Juarez*, which belonged to the 308th Bomb Group. *Hill*

*Rangy Lil*, a B-24L, serial 44-41867, was assigned to the 528th Bomb Squadron, 380th Bomb Group. This aircraft was among those sent for recycling in the U.S. at the RFC Smelt Depot 31 October 1945. *Crow*

*"QUEEN OF THE STRIP"*, a B-24L, serial 44-42456, was assigned to the 529th Bomb Squadron, 380th Bomb Group. In one hand the queen holds her crown, while with the other she appears to be waving. This view of the artist-at-work reveals the pains-taking detail that went into the art. *Crow*

*MAID in The USA*, a B-24L, serial 44-40776, was assigned to the 529th Bomb Squadron, 380th Bomb Group. *Campbell*

*Sandra Kay*, a B-24J, serial 42-72790, was assigned to the 530th Bomb Squadron, 380th Bomb Group. This aircraft had its own guardian angel riding on a cloud. *Sandra Kay* was salvaged 19 April 1945. *Crow*

*PLEASURE BENT*, a B-24J, serial 42-73440, was assigned to the 530th Bomb Squadron, 380th Bomb Group. It was salvaged 16 December 1944. *Malcom*

This is no ordinary hobo portrayed on the B-24, serial 44-42201. He wears an American Legion badge and carries the G.I. Bill of Rights among his belongings. *Bread Line in '49* is among the paintings by Ebbeson. This aircraft belonged to the 528th Bomb Squadron, 380th Bomb Group. The crinkled metal and twisted prop suggests the aircraft had a landing gear problem. *Acord*

*PUSS 'N BOOTS*, a B-24J, serial 42-72942, was assigned to the 380th Bomb Group. *Crow*

*Say UNCLE*, a B-24D-70-CO, serial 42-40534, was assigned to the 528th Bomb Squadron, 380th Bomb Group. The aircraft was also known as *THE ONE* after it transferred to the 90th Bomb Group. Note the first monkey's head is that of Tojo while the second is a Yank with a tight grip on his adversary. *Smith*

*"Betty J"* was among the B-24s which served in the Asiatic-Pacific Theater. *Crow*

*SCREAMIN' MIMI*, a B-24D, serial 42-40997, was assigned to the 565th Bomb Squadron, 389th Bomb Group. She was listed as missing in action 28 April 1944. The Germans restored her and she then flew for the enemy in her new assignment in KG 200. During this time the aircraft was shot down while trying to infiltrate a U.S. bomber formation. *Ethell*

*DOUBLE TROUBLE* appears to refer to the serial number of this B-24. The ribbons held by the two females both have the number 100 on them. The serial number of this aircraft was 42-100100. *Double Trouble* flew with the 578th Bomb Squadron, 392nd Bomb Group. *Ethell*

*SHORT-SNORTER*, with the winged horse, *Pegasus*, painted on the side, was one of the B-24s assigned to the 579th Bomb Squadron, 392nd Bomb Group. Its serial was 42-99990. *Ethell*

*OUR GAL*, a B-24J, serial 42-100308, was assigned to the 578th Bomb Squadron, 392nd Bomb Group. *Ethell*

*BAIL-OUT BELLE*, a B-24J, serial 42-72951, was assigned to the 529th Bomb Squadron, 380th Bomb Group. This aircraft was a 100+ mission ship. It was salvaged on 1 August 1945.

The B-24, serial 42-51105, *the SWEETEST (ROSE) OF TEXAS*, flew with the 701st Bomb Squadron, 445th Bomb Group, Eighth Air Force. After completing her service she was returned to the Zone of the Interior. *Ethell*

*STAR DUST*, a B-24, serial 42-7576, was assigned to the 700th Bomb Squadron, 445th Bomb Group. It later transferred to the 446th. *STAR DUST* was retired to the states after completing its military service. *Crow*

*DESPERATE DESMOND*, a B-24, serial 42-7498, was assigned to the 706th Bomb Squadron, 446th Bomb Group. The art is an old time villain complete with a handlebar mustache, in the form of a B-24 carrying two bombs. *Lobotzke*

*'BACK TO THE SACK'* depicts Donald Duck in his night shirt and cap ready to retire for the night. This B-24J, serial 42-51288, was assigned to the 713th Bomb Squadron, 448th Bomb Group. The photo was taken while the aircraft was at Base Air Depot #2, England. *Lobotzke*

The B-24J, serial 42-100109, *BLACK WIDOW*, was assigned to the 713th Bomb Squadron, 448th Bomb Group. It was listed missing in action on 1 April 1944. *Ethell*

*BOOT in the ASS*, a B-24J, serial 42-51369, was assigned to the 724th Bomb Squadron, 451st Bomb Group. *Hill*

A DOG'S LIFE, a B-24J-CO, serial 44-40281, was assigned to the 753rd Bomb Squadron, 458th Bomb Group. The art on this bare-metal Liberator portrays man's best friend playing with a bomb as though it were a huge bone. *A DOG'S LIFE* was one of the B-24s which was originally equipped to drop the Azon bomb. *A DOG'S LIFE* met its demise 13 February 1945 when it crashed near Spixworth. *Hetzel*

*BIG CHIEF lil' Beaver*, a B-24J, serial 42-51514, flew with the 458th Bomb Group. This picture was taken at Watton Air Depot by Sgt. Arthur Lobotzke while it under went armor modifications. *Lobotzke*

*COOKIE*, a B-24, serial 42-95165, flew with the 752nd Bomb Squadron, 458th Bomb Group. *COOKIE* crashed in Rheims, France on 21 January 1945. *Hetzel*

*BABY SHOES*, #555, was a B-24J, serial 42-50555, which flew with the 753rd Bomb Squadron, 458th Bomb Group. Above the "triple nickel" is one of the many variants seen in the forward observer's window. Note the bomb pointing to the bombardier's station next to the name, "TUBBY" Thompson. *Baby Shoes* was retired back to the states after her service was complete. *Hetzel*

*BACHELOR'S BEDLAM*, a B-24H, serial 44-40287, belonging to the 753rd Bomb Squadron, 458th Bomb Group, flaunts a female clothed in only a grass skirt. This B-24 was equipped with Azon antennas for radio-directed glide bombs, a predecessor of today's smart bombs. At the end of service, it was returned to the ZI. *Hetzel*

*FINAL APPROACH*, a B-24J, serial 42-52457, belonging to the 752nd Bomb Squadron, 458th Bomb Group, sports nose art showing the Continental USA. Inside an arrow pointing home, is a caricature of a B-24 with the 458th Group designator, a "K" inside a circle. This o/d B-24 never made that final approach. On 8 April 1945, while on its 123rd mission, *FINAL APPROACH* was hit by flak and crashed in Lechfeld, Germany. *Hetzel*

*"FLAK MAGNET"*, a B-24H-10-CF, serial 41-29273, was assigned to the 753rd Bomb Squadron, 458th Bomb Group. Like most of the groups, the 458th had its shark mouths. *"FLAK MAGNET"* flashed a grimacing smile in which the bombardier perched while lining up on the day's target. *"FLAK MAGNET"* flew its last mission on 22 April 1944, which ended when the aircraft was hit by flak on a mission to Hamm, Germany and crashed at Hoetmar. As happened very often after the loss of a prized aircraft, another aircraft took the name of the "smiling shark." *"FLAK MAGNET II"*, serial 41-28962 was later transferred to the 47th Bomb Group. *Hetzel*

*HERE I GO AGAIN*, a B-24J, serial 42-95179, was assigned to the 752nd Bomb Squadron, 458th Bomb Group. This shiny-bare metal bird shows the comic strip character, *Dagwood Bumstead*, well prepared for the mission. He is carrying all of his flight gear and the famous multi-layered "Dagwood Sandwich." *Hetzel*

*Above right: HOOKEM COW*, a B-24H, serial 42-95120, was assigned to the 754th Bomb Squadron, 458th Bomb Group. *HOOKEM COW* crashed on take-off 14 April 1944. *Hetzel*

*JOLLY ROGER*, a B-24M, serial 44-40475, was a member of the 458th Bomb Group, 8th Air Force. *Hetzel*

*Miss Used*, a B-24L, serial 44-40278, was assigned to the 458th Bomb Group. *Hetzel*

*Liberty "LIB"*, an o/d B-24H-10-CF, serial 41-29303, flew with the 752nd Bomb Squadron, 458th Bomb Group and sports the symbol of our liberty. The name was a combination of the statue and the aircraft type, Liberator. *Hetzel*

*"LASSIE COME HOME"*, a B-24J-155-CO, serial 44-40283, was assigned to the 753rd Bomb Squadron, 458th Bomb Group. It was another of the few B-24s assigned to the 458th equipped with the Azon antenna. After a mission on 14 January 1945 to Hallendorf, *"LASSIE COME HOME"* crashed on approach to Horsham at Miles Cross. *Hetzel*

*"PRINCESS PAT,"* a B-24J, serial 42-95316, was assigned to the 752nd Bomb Squadron, 458th Bomb Group. It crashed 9 April 1944 at Horsham. *Hetzel*

*Silver Chief*, a B-24J-150-CO, serial 44-40201, flew for the 753rd Bomb Squadron, 458th Bomb Group. It was salvaged 11 January 1945 at Horsham. *Hetzel*

*YOU CAN'T TAKE IT WITH YOU!*, a B-24J, was assigned to the 458th Bomb Group. The young lady is holding a string with a cherry attached to the end. *Hetzel*

*"TOGGLE ANNIE"* was one of the B-24s assigned to the 464th Bomb Group, Fifteenth Air Force. The female in this painting, like so many others, is wearing an added almost invisible dress probably due to orders from the C/O. *Moffitt*

*"SKY ROOM"*, a B-24J, was assigned to the 754th Bomb Squadron, 458th Bomb Group. The art portrays a flying out house with a B-24 body. On the flag which flies over the door of the out house is written, "NO GUM CHUMB!" on the side in white are the words, "POSITION UNCERTAIN." There is a sign stuck on the bottom of the door which says, "HILLBILLY SMOKING TOBACCO 3 CENTS." To finish the motif the positions of both pilot and co-pilot are noted at the very bottom of the door.*Hetzel*

This shark mouth had its picture made at Watton Base Air Depot #2 in England. This photo and others of the 458th were taken by Sgt. Arthur Lobotzke on their way through the depot. *Lobotzke*

*Spotted Ass Ape*, a B-24D, serial 41-28967, was assigned to the 754th Bomb Squadron, 458th Bomb Group. This war-weary formatter not only had the spotted body for high visibility, but it also had a wonderful shark mouth design on the nose giving it the appearance of a large spotted shark. *Hetzel*

*OPEN POST*, a B-24J, serial 42-50449, was assigned to the 753rd Bomb Squadron, 458th Bomb Group. Its flying career was ended by a mid-air collision 23 February 1945 over Felmingham. Photo taken at Watton Depot. *Campbell*

*The "TRAVLIN" BAG*, a B-24J, serial 42-50912, was assigned to the 753rd Bomb Squadron, 458th Bomb Group. *Hetzel*

*Heavenly Hideaway*, a B-24J, serial 44-40277, was originally named *Miss Used*. It had flown as a member of the 753rd Bomb Squadron, 458th Bomb Group, which was based at Horsham St. Faith Airfield, Norwich, England. It was one of the original Azon aircraft. The aircraft and crew later transferred to the 755th Bomb Squadron which served in the 458th Bomb Group. *Hetzel*

Many of the wonderful pictures of the 458th Bomb Group were taken by Sgt. Herman Hetzel, the group photographer, seen here with one of the group's shark mouth B-24s. Not only did he take the pictures but he also persevered them. At the end of hostilities, Sgt. Hetzel was told by his commander they would have no further need of the photos. He was told to destroy them, but like a true lover of the planes and their art, he tossed his belongings and filled his duffle bag with negatives. *Hetzel*

*POM POM EXPRESS*, a B-24M-1-FO, serial 44-50396, flew with the 531st Bomb Squadron, 380th Bomb Group. This aircraft was one of many which made their way to the states only to end up at the airplane graveyards and smelters such as Kingman, Arizona. *Acord*

*GOLDEN LADY*, a B-24D, serial 42-40914, was assigned to the 321st Bomb Squadron, 90th Bomb Group. The gold tinted female appeared on both sides of this aircraft which was named for her. Along with the many gold bomb mission markers seen on this ninety-seven plus mission aircraft are six gold Japanese Zero markers. The first was for an attacking Zero which had put a 20mm shell through the plane before being shot down. The remaining five were earned in a daylight mission over Rabaul. *Horton*

Running to escape losing his head for the Japanese Empire, this little guy was the center of the art on *Nipponese Clipper*, a B-24, serial 42-72966, which flew with the 38th Bomb Squadron, 30th Bomb Group. *Horton*

*NOCTURNAL MISSION*, a B-24J-205-CO, serial 44-41263, was assigned to the 529th Bomb Squadron, 380th Bomb Group. After it completed combat service, this aircraft was sent to Kingman, Arizona to the smelter. *Acord*

*Shy-Chi Baby*, a B-24J, serial 44-40920, was one of the Raven aircraft belonging to the 531st Bomb Squadron, 380th Bomb Group. *Shy-Chi Baby* fell short of reaching the century mark by only three missions. *Acord*

*Flak Fled Flapper*, a B-24J-CO, serial 44-40434, was assigned to the 528th Bomb Squadron, 380th Bomb Group. After completing 1,000+ hours, this aircraft was salvaged on 28 September 1945 on Biak. The art on *Flak Fled Flapper*, like many of the 380th's nose art, was painted by Ebbeson. *Horton*

*Undecided*, a B-24J-125-CO, serial 42-109990, was one of the aircraft flown by the 530th Bomb Squadron, 380th Bomb Group. *Undecided* was sent to the Kingman smelter in November 1945. *Acord*

*MY DEVOTION*, a B-24M, serial 44-42248, was a member of the 528th Bomb Squadron, 380th Bomb Group. *Acord*

*"QUEEN OF THE STRIP"*, a B-24M-35-CO, serial 44-42456, was assigned to the 529th Bomb Squadron, 380th Bomb Group. *Acord*

*TEXAS KATE*, a B-24J CO, serial 44-40358, was assigned to the 392nd Bomb Squadron, 30th Bomb Group. It was piloted by Capt. H.S. Trewitt who's home town was Dallas. Beside his name is written the phrase, *Another "TEXAS AGGIE" '37 Kate's Ol Man*. The crew chief was M/Sgt. J.H. Springer. His name is followed by the name *Joaquin*. *Acord*

*PEACE OFFERING*, a B-24L-20-FO, serial 44-50250, was assigned to the 529th Bomb Squadron, 380th Bomb Group. *Acord*

*MOONLIGHT MAID*, a B-24, serial 44-42331, was assigned to the 529th Bomb Squadron, 380th Bomb Group. *Acord*

*"MILADY"*, a B-24, serial 42-73134, was assigned to the 531st Bomb Squadron, 380th Bomb Group. This secretary came complete with pencil and pad for taking dictation while riding a bomb. *"MILADY"* was lost, along with six members of the crew, in a training exercise 17 January, 1945. *Horton*

*STATE SIDE STUFF*, a B-24, serial 44-51301, belonged to the 530th Bomb Squadron, 380th Bomb Group. *Horton*

The men who crewed this aircraft were, like all of the others, hunters. The art on the B-24, *WOLF PACK*, shows them during their preferred hunt. *Horton*

*SHOO-SHOO BABY*, a B-24J, serial 43-109984, was assigned to the 408th Bomb Squadron, 22nd Bomb Group in the Southwest Pacific. This aircraft was named for a popular song of the time. *Horton*

FIRE
EXTINGUISHER

*Playmate*, a B-24, serial 44-40791, was assigned to the 867th Bomb Squadron, 494th Bomb Group. It was returned to the states upon the completion of its service. *Davis*

*Hay Maker*, a B-24, serial 44-40725, transferred from the 867th Bomb Squadron, 494th Bomb Group, on loan to the 319th Bomb Squadron, 90th Bomb Group where this photo was taken. While on loan the aircraft was shot down by fighters on a mission to Balikpapan on 30 September 1944. The crew was lost. *Horton*

The B-24, *Star Eyes*, was assigned to the 22nd Bomb Group, Fifth Air Force. *Horton*

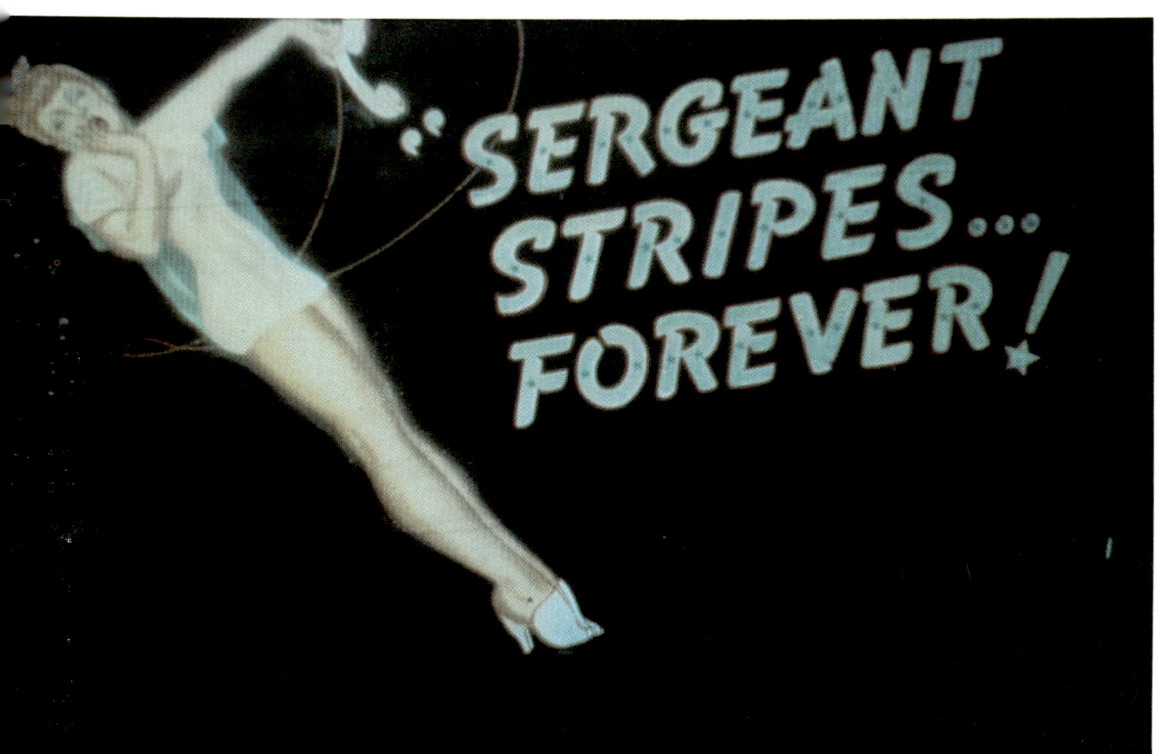


*"SERGEANT STRIPES. . . FOREVER!"* was among those B-24s which received their art at Fairfield Suisan, now Travis Air Force Base, before being sent overseas to face combat. *Horton*

HAM
RAZZL

*RAZZLE-DAZZLE!*, an O/D B-24, was adorned with the female gremlin, *Fifinella*, in honor of the Women's Air Service Pilots (WASPs) who ferried all types of aircraft from the U.S. to their respective destinations. The art was applied at Fairfield Suisan, now Travis Air Force Base, before the aircraft's deployment. *Horton*

A very elaborate paint scheme went on this B-24 assigned to the 64th Bomb Squadron, 43rd Bomb Group named *WINDY CITY KITTY*. *Horton*

*"Out of This World"*, a B-24J, serial 44-40681, was assigned to the 819th Bomb Squadron, 30th Bomb Group. *Horton*

*WELL DEVELOPED*, an F-7, serial 44-40209, was assigned to the 2nd Photo Mapping Squadron in the Southwest Pacific. *Shaw, via Hernandez*

*"LONG DISTANCE"* was another of the B-24s decorated for the awaiting crews while at Fairfield Suisan. *Horton*

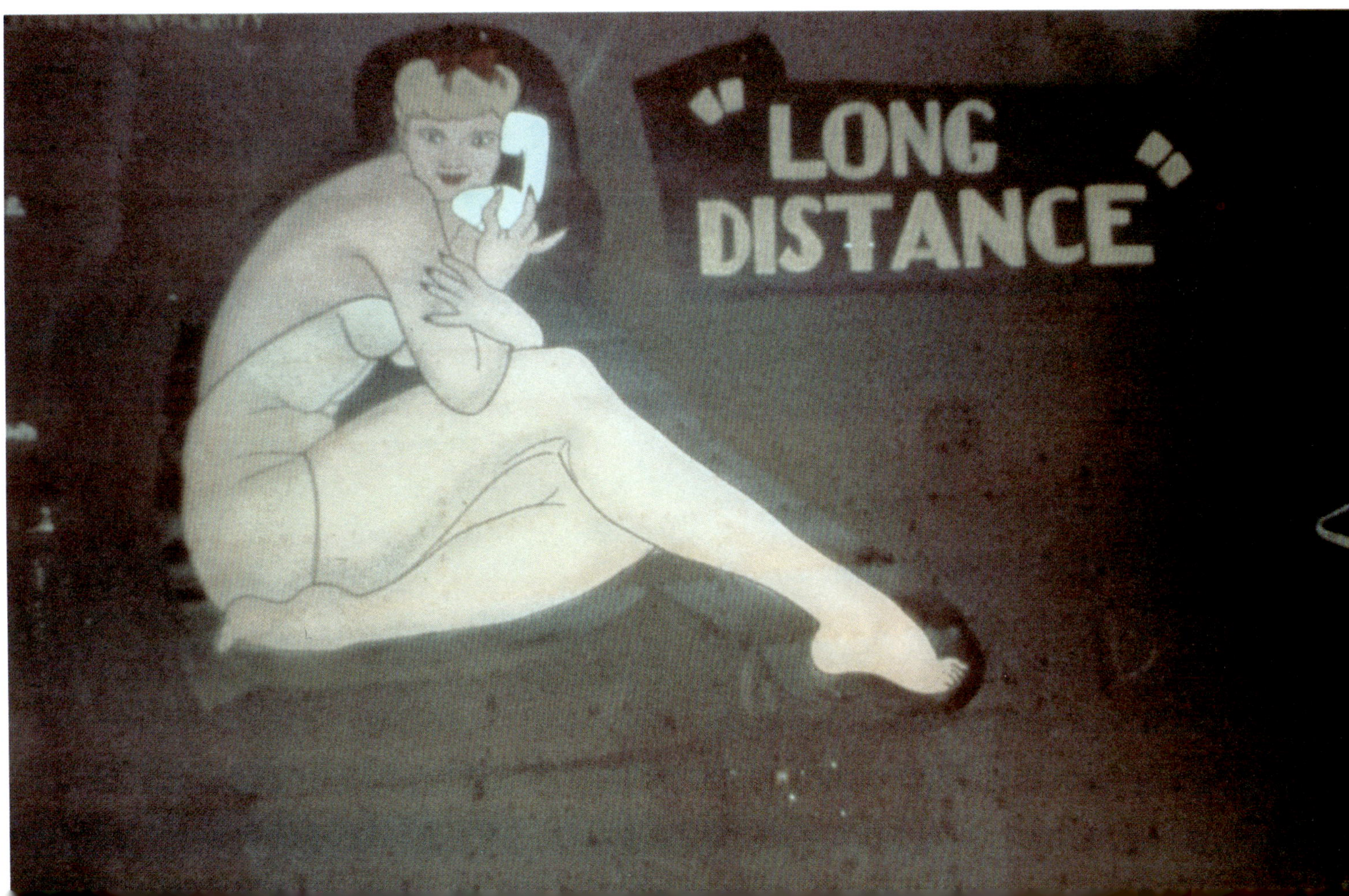

This photo of the B-24, *"SWEET SIXTEEN,"* was taken at Fairfield Suisan before deployment overseas. *Horton*

*"I'LL BE AROUND"* also had its picture taken at Fairfield Suisan after the art was applied. *Horton*

*The FLAKMAN* was on a B-24 belonging to the 464th Bomb Group. Proud of its victories against German fighters, the crew showed seven large swastikas on the side of their aircraft under the many bomb mission markers. *Ethell*

*VIRGIN STURGEON/"NEEDS ON URGIN"*, a B-24H, serial 42-52516, was assigned to the 784th Bomb Squadron, 466th bomb Group. The "V" on virgin is partially covered by the armor plate applied to many 8th Air Force aircraft to afford the pilot and co-pilot added protection. *Beitling*

*OFF LIMITS AGAIN*, a B-24J, serial 42-94974, was assigned to the 786th Bomb Squadron, 466th Bomb Group. Here the protective armor was also used as a score board showing thirteen bomb missions and three Luftwaffe fighters shot down. *Beitling*

*"BETTA DUCK"*, a B-24J, serial 44-40454, was a member of the 785th Bomb Squadron, 466th Bomb Group. This picture taken at Base Air Depot #2. *Lobotzke*

*TIMES A' WASTIN*, a B-24J, serial 42-50569, was a member of the 758th Bomb Squadron, 466th Bomb Group. A jug-toting Snuffy is running in the direction of a road sign which reads "BERLIN BREWERY 10 MILES." In keeping with the art, the score board is titled, *SNUFFY'S PUBING MISSIONS*. *Campbell*

*The "MAD MONK!"*, a B-24, serial 41-29402, was assigned to the 786th Bomb Squadron, 466th Bomb Group. *Ethell*

Donald Duck in tails and top hat accompanied the name *PLAY BOY* on this B-24J, serial 41-29399, which was assigned to the 784th Bomb Squadron, 466th Bomb Group. Donald appears to be thinking and next to him is written, "NO BULLET PROOF GLASS." *Play Boy* was listed as missing in action 29 April 1944 when it crash landed in Daarle, Holland after a bombing raid on Berlin. *Ethell*

*BONNIE*, a B-24J, serial 41-29459, was assigned to the 784th Bomb Squadron, 466th Bomb Group. *Ethell*

*Jamaica?* was an adaptation of a question asked by members of the crew when fellow members returned from dates. This 466th Bomb Group B-24J, serial 41-28746, served with the 785th Bomb Squadron. It was reported missing in action 25 September 1944. *Bietling*

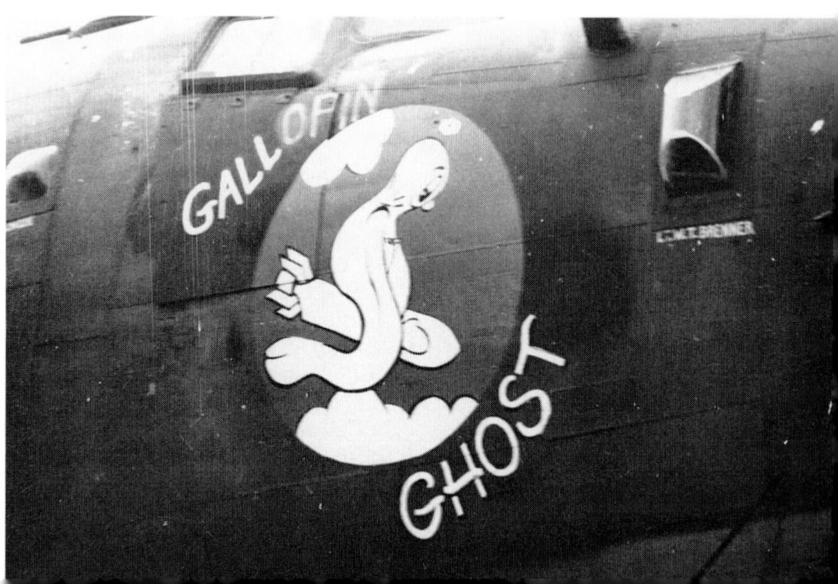


*GALLOPIN' GHOST*, a B-24J, serial 41-29439, was assigned to the 789th Bomb Squadron, 466th Bomb Group. It crashed 29 September 1944 on a trucking mission. *Bietling*

*QUEEN OF HEARTS*, a B-24H, serial 42-52511, was assigned to the 786th Bomb Squadron, 466th Bomb Group. It met its end in a crash-landing at New Romney, Kent after a mission to V-weapons site on 17 July 1944. Note the little indian riding an arrow on shooting a bomb. *Ethell*

*PARSON'S CHARIOT*, a B-24J, serial 42-110162, was a member of the 784th Bomb Squadron, 466th Bomb Group. It was salvaged 3 April 1945. *Bietling*

*TAIL WIND*, a B-24H, serial 41-29368, was assigned to the 789th Bomb Squadron, 467th Bomb Group, Eighth Air Force. *Bietling*

The B-24J, serial 44-40070, was named *MADEMOISELLE ZIG ZIG*. The art on this 790th Bomb Squadron, 467th Bomb Group aircraft, showed a stocking clad female in the early stages of dressing. *Bietling*

*"Blonde Bomber,"* a B-24J, serial 42-50471, was assigned to the 467th Bomb Group. It later transferred to the 453rd and from there was returned to the states. *MAFB*

*Capricorn* was chosen for the B-24, serial 42-52744. *Lobotzke*

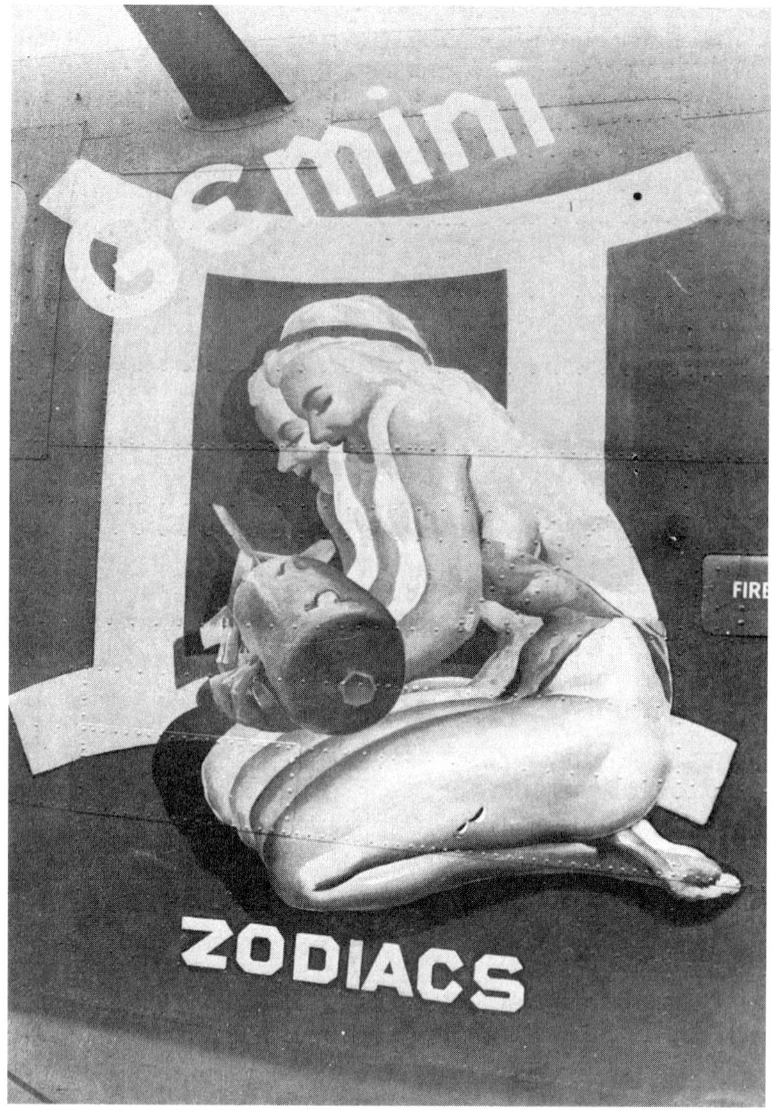


*Gemini*, was also among the B-24s assigned to the 834th Bomb Squadron, 486th Bomb Group. It carried the serial 41-29490. *Hill*

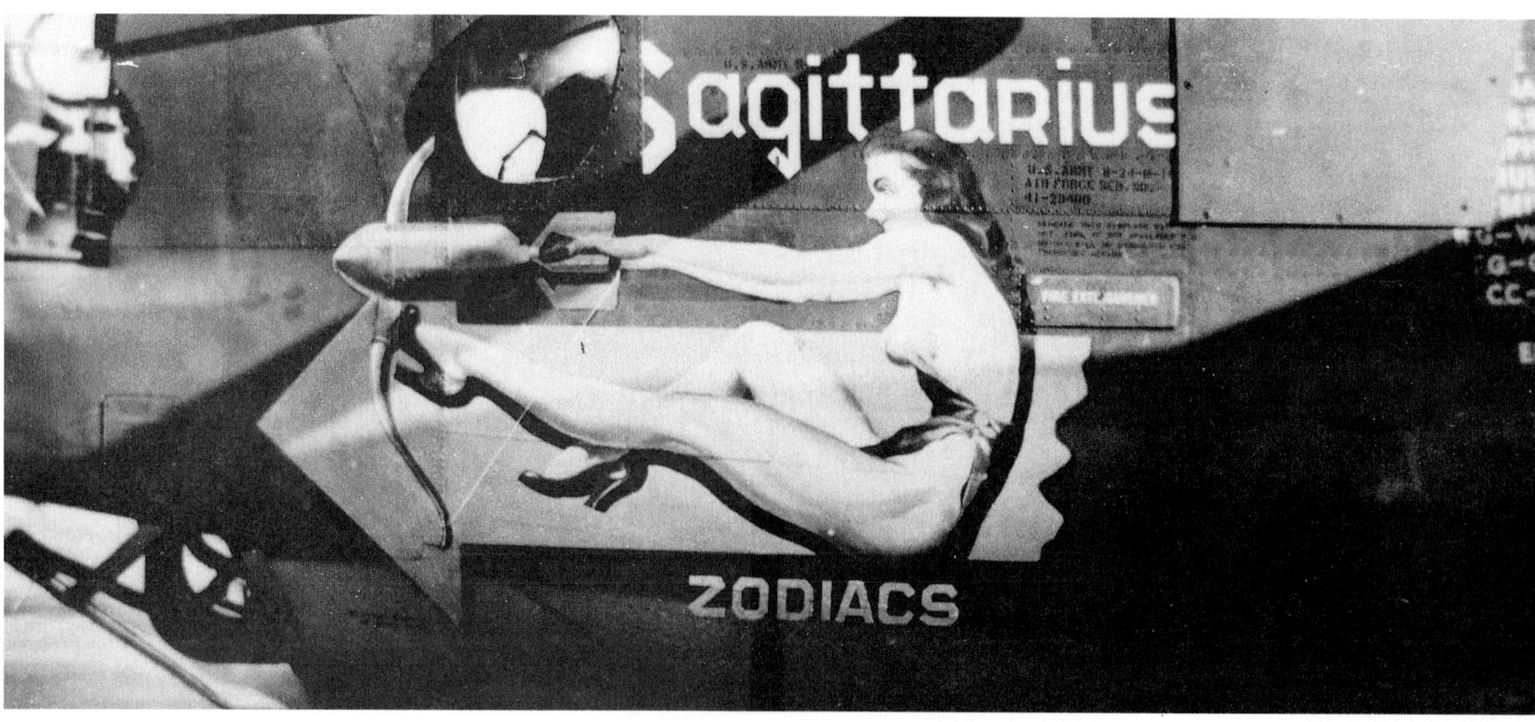


*Sagittarius* was redesigned to show the loading of a bomb replacing the arrow in the drawn bow on the B-24, serial 41-29400. *Hill*

*AQUARIA*, was the name given to the art chosen for the B-24, serial 42-52545. *Hill*

*Libra* was selected for the B-24, serial 42-52508. *Lobotzke*

The 834th Bomb Squadron, assigned to the 486th Bomb Group, elected to use the signs of the zodiac, painted by Phil Brinkman, as the theme of the art on their B-24s. *Aries* was on the B-24, serial 42-52765. *Lobotzke*

*Leo* was on the B-24, serial 41-29605. *Aries, Libra, Capricorn and Leo* were all captured by the camera of Sgt. Arthur Lobotzke while they were at Base Air Depot #2, England. *Lobotzke*

*"PURTY BABY,"* a B-24H, serial 44-40299, was assigned to the 487th Bomb Group. It was said by one of the soldiers, "After a look at her, anybody would smile, look away, and dream of home . . . Reet?" This art was the creation of Sgt. Duane Bryers. *Lobotzke*

*YOU'RE SAFE AT HOME*, a B-24H, serial 42-94827, flew with the 851st Bomb Squadron, 490th Bomb Group. The 490th attacked interdictory targets and supported advancing ground forces after the Battle of the Bulge. *Lobotzke*

*LITTLE IRON PANTS*, a B-24J, serial 42-94781, was assigned to the 849th Bomb Squadron, 490th Bomb Group. This little one is carrying a club that has a spike sticking through the end. *Lobotzke*

*OLE BALDY*, a B-24J, serial 42-94863, was assigned to the 851st Bomb Squadron, 490th Bomb Group. The bald eagle, which stands for everything American, is getting involved by dropping a bomb in this piece of art. *Lobotzke*

*"DORTY TREEK,"* a B-24H, serial 44-10485, was assigned to the 853rd Bomb Squadron, 491st Bomb Group. *Lobotzke*

*"Lucy Quipment,"* a B-24H, serial 41-29545, was assigned to the 851st Bomb Squadron, 490th Bomb Group. It met its end when it crashed on 17 July 1944. *Bietling*

*CALL ME "Savage,"* serial 42-94937, was one of the B-24s belonging to the 849th Bomb Squadron, 490th Bomb Group. *Ethell*

*"HOUSE OF RUMOR,"* was on a B-24J, serial 44-40271, belonging to the 854th Bomb Squadron, 491st Bomb Group in North Pickenham, England. It was shot down by German fighters 26 November 1944. This photo was taken at Base Air Depot #2 in 1945. *Lobotzke*

Bugs Bunny was the man of the hour on the B-24J, serial 42-110107, *"WHAM! BAM! THANK YOU, MA'M."* This aircraft, belonging to the 854th Bomb Squadron, 491st Bomb Group, got its picture taken at Base Air Depot #2. *Lobotzke*

*THE FLYING JACKASS*, a B-24H, serial 44-40239, featured a jackass carrying members of the crew. This was one of the aircraft belonging to the 853rd Bomb Squadron, 491st Bomb Group. *Hill*

*"What's Cookin' Doc?"* was the question carried by this painting of a Super Bugs Bunny. It was found on the B-24, 42-110157, which was assigned to the 855th Bomb Squadron, 491st Bomb Group. The aircraft later transferred to the 466th Bomb Group before its return to the states. *Hill*

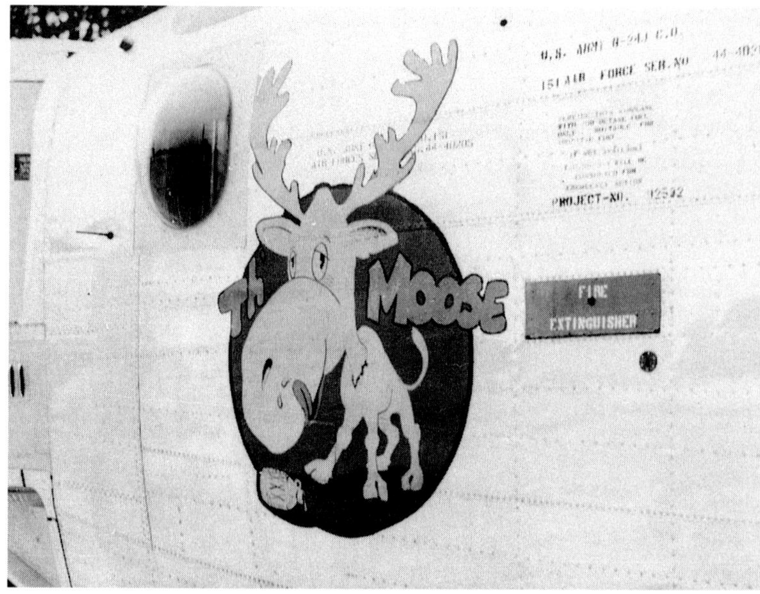


*The MOOSE*, serial 44-40205, was one of the B-24s assigned to the 853rd Bomb Squadron, 491st Bomb Group. It was reported missing in action 26 November 1944. *Hill*

The B-24, *The LATRINE RUMOR*, serial 44-40271, was assigned to the 854th Bomb Squadron, 491st Bomb Group. It was among those which visited Base Air Depot #2. This aircraft was listed as missing in action 26 November 1944. *Lobotzke*

*SCREW BALL*, a B-24H, was used in the *Carpetbagger* missions. Its black paint scheme helped camouflage it in the night skies as it went in at low levels to drop its cargo. The most unusual thing about the *Carpetbagger* B-24 was its cargo. In place of the usual bomb drops these aircraft transported agents, supplies and propaganda leaflets. *Beitling*

The young lady on *S.O.L.* seems to be up a creek without a paddle. This B-24, serial 44-40066, was assigned to the 858th Bomb Squadron, 492nd Bomb Group, but later transferred to the 753rd Bomb Squadron, 458th Bomb Group. *Ethell*

*"BOTTLE BABY,"* a B-24J, serial 44-40169, was assigned to the 858th Bomb Squadron, 492nd Bomb Group. The name is accompanied by the painting of a bomb-carrying female riding a bottle bare-back. *Ethell*

*THE SNIFFIN GRIFFIN*, a B-24J, serial 44-40705, was assigned to the 865th Bomb Squadron, 494th Bomb Group. It was returned to the states after completing its tour of duty. *MAFB*

The B-24J, *Old Sack*, serial 44-40321, flew with the 493rd Bomb Group as they fought Hitler in his own backyard. This beach bunny is the art of Salega. *Ethell*

With only the name *Merchant* written, the lady holding the fan of feathers completes the name of this B-24 which was assigned to the 493rd Bomb Group. This photo of the *Feather Merchant* was taken at Base Air Depot #2 in England. *Labotzke*

*HAIRLESS JOE*, a B-24J, serial 44-40437, was assigned to the 860th Bomb Squadron, 493rd Bomb Group. This mountain man carries a club in his right hand and a jug of moonshine in his left. *Lobotzke*

*Left: "RIOT CALL,"* a B-24, serial 44-40755, was assigned to the 494th Bomb Group. The aircraft was returned to the states after its service was completed. *Crow*

*Right: SUPERCHICK*, a B-24L, serial 44-40761, was assigned to the 865th Bomb Squadron, 494th Bomb Group. She was sent to the smelter ovens 31 December 45. *Crow*

*Left: SACK TIME SAL II*, a B-24, serial 44-40796, was assigned to the 866th Bomb Squadron, 494th Bomb Group. She was retired to the Zone of the Interior. *Crow*

*Right: STAR DUST*, a B-24L, serial 44-41610, was assigned to the 867th Bomb Squadron, 494th Bomb Group. After the war ended *Star Dust* was returned to the states. *Crow*

*Left: Les Miserables*, a B-24L, serial 44-40666, was assigned to the 866th Bomb Squadron, 494th Bomb Group. The 866th's squadron insignia was a black chimpanzee wearing flight gear and carrying a machine gun. This was the bases of the art showing two chimps on a flying carpet sporting the national insignia. *Les Miserables* was reported missing in action 9 October 45. *Crow*

*Right: PACIFIC PASSION*, a B-24L, was a Head Quarter's plane assigned to the 494th Bomb Group. *Crow*

*CONTRARY MARY!*, a B-24J, serial 44-40739, was a member of the 864th Bomb Squadron, 494th Bomb Group. This aircraft retired to the states after its service. *Ethell*

*SABU/"FLYING BOXCAR"* was a B-24J, serial 44-40749, which flew with the 867th Bomb Squadron, 494th Bomb Group. *Hill*

The B-24J, serial 44-40286, *THE GREEN HORNET*, was assigned to the 493rd Bomb Group. *Hill*

There were many *Shoo Shoo Babys* adorning the aircraft of World War II. This one was on a B-24J, serial 42-50284, in the 8th Air Force. *Lobotzke*

*BUZZ TAIL II*, a B-24, serial 42-50447, shows a rattle snake coiled around a falling bomb. The picture was taken at Base Air Depot #2, England. *Lobotzke*

Riding a bomb on its way to Berlin, according to the sign, is a popular cartoon character, Pop-Eye the Sailor. The title of the art on this B-24 is *POOP DECK PAPPY. Lobotzke*

The term *4-F* appeared on many different types of aircraft in different locations throughout the war. This one was on a B-24 which went through the Base Air Depot #2, England. The art was the grim reaper with a vulture on his shoulder. *Lobotzke*

*"THIS ABOVE ALL,"* a B-24J, serial 44-40328, was assigned to the 493rd Bomb Group. The picture was taken at Base Air Depot #2, Watton. *Lobotzke*

There is no name visible on this B-24's art. However, the wrist watch with a seated female is one of the more unique pieces of art. This B-24 was at Base Air Depot #2, England. *Lobotzke*

Several of the artists during World War II used the Liberty Bell, our symbol of freedom, and with an added letter *e* to the *Bell* gave it a different connotation. This B-24L, *Liberty Belle*, was in the hangar at Base Air Depot #2, England. *Lobotzke*

After the bombs have been dropped the only objective of the crew was to return to base. In hopes of completing this objective, several planes were painted with the proverbial, *bad penny*, which always comes backs. The B-24, *The Bad Penny*, was photographed by Sgt. Arthur Lobotzke while he was stationed at Base Air Depot #2, England. *Lobotzke*

*Left:* This young lady, riding a wild bull, was painted on a B-24 which went through Base Air Depot #2 Watton. *Lobotzke*

*Right:* The name *Heavenly Body* on this B-24 must refer to the heights reached by the aircraft, or the launching point of the bombs it dropped. It surely wasn't in reference to the *Big Foot* creature painted on the aircraft. This is among the many B-24s which went through the Base Air Depot #2, England. *Lobotzke*

The B-24H, *Mamma's Kid's*, has the art of Cpl. Henley. It has its own means of depicting the mission of dropping-its-load on the enemy with the words *Bombs Away*. A member of the Eighth Air Force this B-24 was serviced at the Base Air Depot #2, England. *Lobotzke*

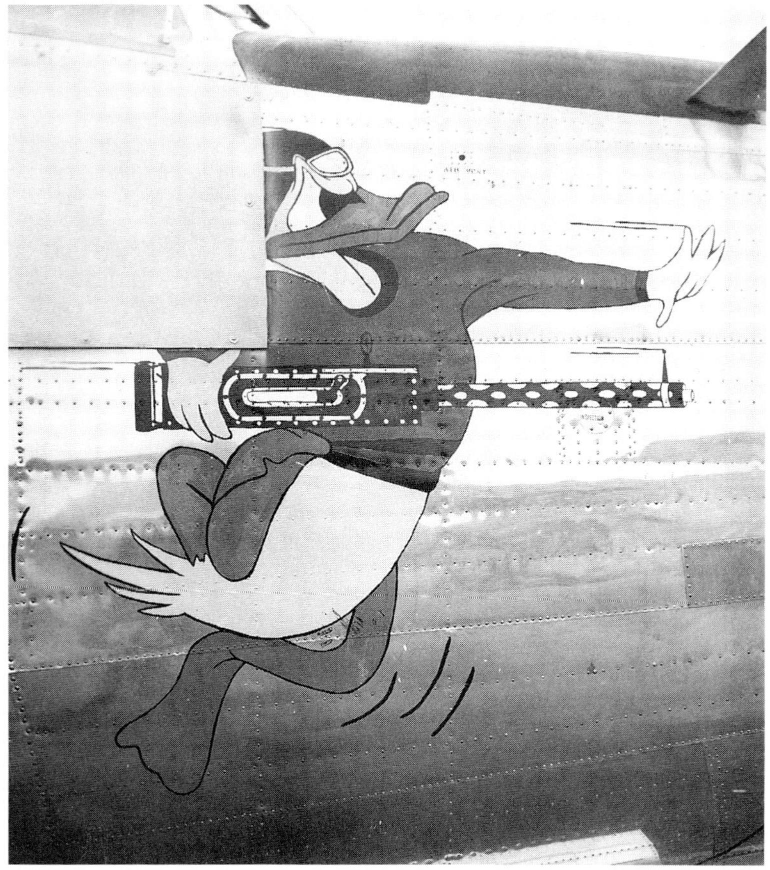

The added protective shield has covered a portion of the art on this B-24 which sports a fighting Donald Duck. This Eighth Air Force B-24 had its photo was taken at Base Air Depot #2, England. *Lobotzke*

Complete with fangs, this hairy wolf walks on two legs as he stalks his innocent prey. This B-24 name *The WOLF PATROL* was at Watton, England. *Lobotzke*

The Eighth Air Force B-24, *The WILD HARE*, featured that crazy Warner Brother's cartoon character, Bugs Bunny. He is heavily armed with a .50 caliber machine gun, a 500lb. bomb, and the very familiar carrot prop. The photo was taken while the aircraft was at Base Air Depot #2, England. *Lobotzke*

Many of the young men who were called on to fight in the skies were more at home on a tractor than in a bomber. Here we have a farmer on his tractor which has been modified with wings and propeller. The little tyke on this B-24 is lending a helping hand pulling-it-through. The photo was taken a Base Air Depot #2, England. *Lobotzke*

The B-24 named *"The THUNDER MUG"* posed for this picture while at Base Air Depot #2, England. *Lobotzke*

The bare-metal B-24L, *BOMBLE BEE*, has a bumblebee which is not only armed in the tail with a set of .50s but is also dropping a bomb over its target. The photo was taken at Base Air Depot #2, England. *Lobotzke*

The art on the B-24J, *Makin' Believe*, features a popular Varga Girl which was on many of the aircraft of World War II. *Makin' Believe* was photographed at Base Air Depot #2, England. *Lobotzke*

*"MAXIE'S" NIGHTMARE* was among the B-24s in the hangar at Base Air Depot #2, England. *Lobotzke*

This Pacific B-24, *SITTING PRETTY*, shows a female elevated to the top of the world. This is a design which was seen often in varied dress and position. *Campbell*

A scantily clothed, bomb-toting female adorns the B-24, *HELL and BACK*. The photo was taken at Base Air Depot #2, England. *Lobotzke*

This B-24 portrays the changing duties of the ladies left in charge of the home front. In her apron and heels, the former homemaker carries a laundry basket of bombs. While still maintaining their femininity, these hard-working women, often referred to as *Ruby the Riveter*, were able to supply the aircraft, ammunition, and supplies necessary to the success of their fighting men. These ladies should not be forgotten. *Malcom*

The B-24, *The Guiding Light*, is an example of some of the artistic talent that was available in the Pacific war zones. *Campbell*

In his tux and top hat, this *Reddy Teddy* adorned one of the B-24s which was serviced at Base Air Depot #2, England. *Lobotzke*

The B-24J, *BLASTED EVENT!*, pulls a switch on *blessed event*, by switching the stork for a duck, which happens to have its tail on fire. This was among the aircraft traveling through Base Air Depot #2, England. *Lobotzke*

BLUE
DREAMS

# Chapter Three

# Flying Fortress

The Boeing B-17 is best known for its rugged dependable airframe that was able to withstand almost any kind of punishment. This aircraft found its fame in becoming the first bomber to complete twenty-five missions in Europe. To this day it is certain to raise arguments whether or not it was the preferred bomber to crew on, during the war it was none the less favored. The stories of missions surviving on just a thread, bear testament to the durability of the "Flying Fortress."

While the B-17 was better known for its service over fortress Europe, it also saw action in the Pacific at the beginning of World War II and in the middle east – even a few served in the desert war.

12,731 B-17s were built: Boeing built 6,981, Douglas Aircraft built 3,000, and 2,700, known as the Vega, were built by Lockheed.

B-17 losses were high, as it went on early war missions with limited fighter escort – approximately 4,750 Flying Fortresses were lost to enemy guns.

Yet the B-17 has won a place in the hearts of the men who flew it and the admiration of many aviation enthusiasts of today.

*BLUE DREAMS*, a B-17G, serial 42-37761, was assigned to the 91st Bomb Group. In the right-hand corner there is a stork carrying a baby, possibly announcing the arrival of a new member in the family of a crewman. *Ethell*

*"HELLS KITCHEN,"* a B-17F, serial 41-29236, was a member of the 44th Bomb Group, Eighth Air Force. Along with its many bomb missions, this aircraft was responsible for destroying eight of Hitler's fighters and one freighter. *AFM*

*MIAMI Clipper*, a B-17F, serial 42-29815, was assigned to the 91st Bomb Group, after its transfer from the 306th 14 September 1943. It then served with the 91st until its return to the states 6 April 1944. The art shows a female with a full moon and palm trees for a back drop. *Snodgrass*

A bolt of lightning sends a startled Hitler running out of an outhouse on this 401st Bomb Squadron B-17F, serial 42-3073, named *LIGHTENING Strikes*. The 91st Bomb Group lost this aircraft, piloted by Lt. William Simmons, on 21 February 1944 when it ditched in the North Sea due to fighter-inflicted damage. *Linsey*

The B-17G, *WICKED WITCH*, was assigned to the 323rd Bomb Squadron, 91st Bomb Group. *Watts*

*MY BABY*, a B-17G, serial 42-107033, piloted by E.R. Kelly, was assigned to the 322nd Bomb Squadron, 91st Bomb Group. *MY BABY* crashed on a mission to Ludwigshafen 5 September 1944 due to mechanical failure. *Watts*

*DUKE OF PADUCAH*, a B-17G, sporting a bare-foot zoot-suiter, was assigned to the 401st Bomb Squadron, 91st Bomb Group. *Watts*

A small percentage of aircraft carried a religious message. This B-17F, serial 42-5712, belonging to the 91st Bomb Group, was named *My Prayer*. The scroll carried this prayer, *"Yea thou I fly through the shadow of the valley of death I fear no evil for thou art with me." MAFB*

Named for its pilot, 1st Lt. Robert Sherriff, the B-17G, serial 42-97151, *SHERIFF'S POSSE*, was one of the aircraft of the 91st Bomb Group. *MAFB*

The art on this B-17F, serial 42-3043, belonging to the 91st Bomb Group, shows Hitler in a precarious situation. The 401st Bomb Squadron's, *HITLER'S GREMLIN*, quotes the Fuhrer as saying, *"Haf a heart,-mine goose-step ain't vot it use to be!* This aircraft, piloted by Lt. Eugene Lockhart, ditched in the North Sea after receiving battle damage on the Schweinfurt/ Regensburg raid on 17 August 1943. *MAFB*

This interesting combination of bomb, bird, and etc. was named *The JUBJUB BIRD* by its crew in reference to the sound made by the B-17's engines. The B-17G, serial 42-31883, belonged to the 401st Bomb Squadron, 91st Bomb Group. *The JUBJUB BIRD'S* service was prematurely ended when it was brought down by flak and crashed near Leipzig. *MAFB*

The comic character, Alley Oop, rides a saber tooth tiger on the nose of the B-17F, *Blooming Grove*. In the background is another B-17 with the same art. It appears to be a group or squadron designation. The 326th Squadron's insignia was Alley Oop astride a dinosaur while the 407th Squadron's insignia showed him riding the back of a bomb-carrying pterodactyl. Both squadrons were members of the 92nd Bomb Group. *MAFB*

This modest piece of nose art on a B-17F makes its point. *Bundles for Berlin* was assigned to the 92nd Bomb Group. *MAFB*

Picturing Hitler's head hanging like a trophy from the small portion of a V, is the B-17F, *Exterminator*. It was assigned to the 92nd Bomb Group. *AFM*

The devil, with his pitchfork aimed downward, sends a definite message to the viewer of the B-17F, serial 42-5070, *HELLZAPOPPIN*. This aircraft was piloted by Capt. Oscar O'Neill, of the 401st Bomb Squadron, 91st Bomb Group. *HELLZAPOPPIN* was shot down by fighters 17 April 1943 and crashed near Nikelausdorf on the Bremen mission. *MAFB*

The crew poses in front of this 92nd Bomb Group B-17F, *CAROL JANE*, serial 42-97121. *Ethell*

*ZOOT BOYS PRIDE*, a B-17F, was assigned to the 401st Bomb Squadron, 91st Bomb Group. *Ethell*

The flak bursts are getting to the well known Disney character, Donald Duck, on the 92nd Bomb Group's B-17F, *FLAK HAPPY. MAFB*

The mascot for Esquire Magazine, *Esky*, rides a bomb on this B-17G, serial 42-31946, piloted by Lt. Donald E. Swanson of the 333rd Bomb Squadron, 94th Bomb Group. Esky was shot down by flak on a mission to Pas De Calais/Cherbourg on 20 April 1944 and crashed near Imotski. *AFM*

*Shackeroo! II*, a B-17F, serial 42-29708, project #92115-R, was assigned to the 332nd Bomb Squadron, 94th Bomb Group, Eighth Air Force. The aircraft was damaged by fighters and ditched in the channel. *McKee*

A very frequently used name, *Double Trouble*, accompanied the two females painted on the 94th Bomb Group's B-17F, serial 42-3082, based at Earls Colne, England. *Essig*

This young lovely dodges an oncoming shell in the artwork named *ROGER DODG'-HER*. This B-17F, serial 42-30425, was assigned to the 333rd Bomb Squadron, 94th Bomb Group, and was piloted by Lt. S.A. Lemly. It was shot down by fighters over Halberstadt and crashed near Abbensen. *McKee*

*VIRGINS DELIGHT*, a B-17G, serial 42-3352, was assigned to the 94th Bomb Group. It was lost in action 30 November 1943 during a mission to Solingen. *MAFB*

*MISCHIEF-MAKER II*, a B-17F, serial 42-30412, was piloted by Lt. Paul Herring of the 339th Bomb Squadron, 96th Bomb Group. On 4 March 1944, while on a mission to Bonn, Dusseldorf, Cologne, Frankfurt, and Berlin, the aircraft fell pray to fighters and was shot down near Vegenstedt. *Huston*

*ACK-ACK ANNIE*, a B-17G, serial 42-32095, was assigned to the 95th Bomb Group. This lovely young thing holds on to her hat as she dodges the flak bursts all around her. *NASM*

*THE KLAP-TRAP II*, a B-17F, serial 42-30130, was assigned to the 337th Bomb Squadron, 96th Bomb Group in Snetterton Heath, England. This aircraft was another of the victims of flak damage and crashed near Thionville on 7 January 1944. *Huston*

*Miss SHAKMATE*, a B-17F, was assigned to the 96th Bomb Group. *Huston*

*"HELL'S" CHARIOT*, a B-17F, serial 42-3442, was assigned to the 96th Bomb Group. Note the bombs spraying from the cluster of German Swastikas. *Huston*

*KIPLING'S ERROR the III*, a B-17F, serial 42-5885, was assigned to the 338th Bomb Squadron, 96th Bomb Group. This aircraft, piloted by Lt. Harry E Johnson, was damaged by fighters and was forced to ditch in the Baltic 11 April 1944. *Huston*

*BLACK HAWK*, a B-17F, was assigned to the 96th Bomb Group. Toward the front of the nose is a reclining female with the name *BETTE LORE* written above her. *Huston*

*FERTILE MYRTLE 3rd*, a B-17F, serial 42-30366, was assigned to the 96th Bomb Group, Eighth Air force. *Fertile Myrtle* carries three swastikas and twelve bomb mission markers. *Huston*

*STINGY*, a B-17G, serial 42-31053, was assigned to the 96th Bomb Group. This photo affords a good view of the Bendix chin turret. The turret was the main distinguishing feature of the G model. *Huston*

*Wabbit-Twacks III*, a B-17F, serial 42-30040, was assigned to the 337th Bomb Squadron, 96th Bomb Group. This bomb-carrying rabbit leaves a foot print on each bomb as he makes his way forward. The art on *Wabbit-Twacks III* was by Sgt. Johnnie White. *Wabbit-Twacks III* was shot down by fighters on the Schweinfurt mission 14 October 1943. *Huston*

*BIG DICK*, a B-17F, serial 42-30123, was assigned to the 96th Bomb Group. Note the girl-art on the door with the name *LITTLE JO*. Note also the artwork on the back of the crew members jacket which is a pair of dice and the name *BIG DICK* matching his aircraft. This aircraft of the 338th Bomb Squadron, piloted by 2nd Lt. David Jarrett was shot down by fighters and crashed near Regensburg on 24 February 1944 on the Regensburg, Augsburg, Stuttgart and Furth raid. *Huston*

The ill fated B-17G, *GOERING'S NIGHTMARE*, was assigned to the 96th Bomb Group. *MAFB*

Although *GOERING'S NIGHTMARE*, met with misfortune the original pattern for the artwork was preserved in this photo. This is another of the works of Johnnie White. He also painted *Wabbit-Twacks III*. *Ethell*

*SUPERSTITIOUS AL-O-YSIS*, a B-17F, serial 42-37807, was assigned to the 100th Bomb Group. At the time of this photo, this aircraft had been on six bomb missions during which they shot down a German aircraft. *Ethell*

Although many aircraft transferred between groups, this B-17F must hold a record. *WABASH CANNONBALL* served with the 91st, 100th, 300th, and 388th Bomb Groups without being shot down. *MAFB*

The picture of the tail of *OLD 494*, a B-17, assigned to the 348th Bomb Squadron, 99th Bomb Group, was taken before her untimely end. The ship was on its 50th mission when it was damaged over an Italian target. However, it brought the crew back, stayed aloft while all parachuted safely to earth, and then was allowed to crash. *MAFB*

Standing in front of the B-17G, serial 42-31066, *Fools Rush In*, are members of the 100th Bomb Group. Note the bomb mission markers on their flight jackets. *Ethell*

*Nine Little Yanks and a Jerk*, a B-17F, serial 42-3271, was a member of the 100th Bomb Group, Eighth Air Force. It has its own female patriot riding a bomb and wearing Uncle Sam's top hat. *Ethell*

This female pose was most often seen in an horizontal position, but it works this way also. *Holy Terror III*, a B-17G, serial 42-40056, of the 351st Bomb Squadron, 100th Bomb Group, was piloted by Lt. Norman L. Chapman. During a mission to Berlin, it was attacked and shot down by fighters. *Ethell*

*"S" For Sugar*, a B-17E, serial 41-24619, was piloted by 2nd Lt. Thomas L. Simmons of the 427th Bomb Squadron, 303rd Bomb Group, Eighth Air Force. The 303rd lost this aircraft to fighters on 11 January 1944, when it crashed near Braunlage. *Rukes*

*SPECIAL DELIVERY*, a B-17G, serial 42-102496, was assigned to the 359th Bomb Squadron, 303rd Bomb Group. *Rukes*

*The FLOOSE*, a B-17G, belonging to the 303rd Bomb Group, started her career in May 1944. Before the end of December the same year, she had piled up an amazing total of 100 missions, without an abort due to mechanical failure. While on loan to another group, *The FLOOSE*, suffering with jammed landing gear, crash landed upon its return to England. The crew members escaped injury. *Rukes*

*Old Crow*, a B-17F, serial 42-31200, was assigned to the 303rd Bomb Group. The crow in this painting is delivering a bottle of his best. *MAFB*

This photo of crew members belonging to the 303rd Bomb Group, in front of the tail section of a B-17F, was taken in early 1943. Note the art on the tail of the ship showing a skeleton prepared to drop a bomb. The vertical stabilizer was a popular area for art in the 303rd. *MAFB*

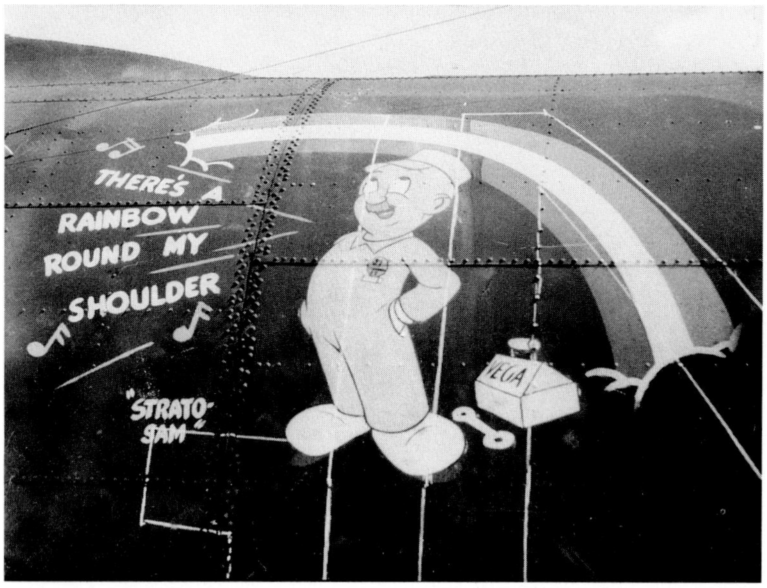


This Vega built B-17 was assigned to the 303rd Bomb Group. It carries the tune, *"THERE'S A RAINBOW ROUND MY SHOULDER." Ethell*

The same character is on this B-17 belonging to the 303rd Bomb Group. This time he is wearing a bunny suit, and announcing, *"WE HAVE SOME EXTRA SPECIAL EGGS FOR THE AXIS THIS EASTER!* Like before this work is signed "STRATO-SAM." *Ethell*

*BAM BAM*, serial 42-37893, a B-17G, was one of the aircraft of the 303rd Bomb Group. *MAFB*

*EARTHQUAKE McGOON*, a B-17G, serial 43-37597, was assigned to the 427th Bomb Squadron, 303rd Bomb Group. This B-17, piloted by Lt. T.F. Kahler, was hit by flak 17 April 1945 on a mission to Dresden and crashed near Brux. *MAFB*

*DUCHESS' DAUGHTER*, a B-17G, serial 42-97272, joined the 303rd Bomb Group on 19 April 1944. At the time of this photo she had been on only four missions. *MAFB*

*SNOOZIN' SUSAN*, a B-17F, serial 42-29860, was assigned to the 351st Bomb Group. *Ethell*

*GREMLIN CASTLE*, a B-17F, serial 42-29850, flew with the 351st Bomb Group. The gremlin is carrying the tools of his trade as he heads for the castle in the distance. Pilots claimed gremlins were collectors of nuts and bolts and often blamed them for mechanical problems. This aircraft was among those at Amarillo Army Air Field. *Snodgrass*

*Target FOR Tonite*, serial 41-24615, served with the 305th Bomb Group. Although she is now retired from combat duty, she carries an impressive score board which shows 75 bomb missions during which the crew shot down 14 German aircraft. This B-17F is parked at Amarillo Army Air Field. *Snodgrass*

The 305TH Bomb Group's B-17F, *DRY MARTINI 4th & THE COCKTAIL KIDS*, had an impressive score board which included twenty German fighters shot down during bomb missions. *MAFB*

One side of the 379th Bomb Group's B-17, *HAG OF HARDERWYK*, showed a nude, horrible, wooden legged female. The hag on this side was quite the opposite. She was a lovely lady holding up a victory sign. This B-17 had the distinction of being the first ship to leave for the combat zones via Dow Field and return to the states via Dow Field. The returning pilots who brought the Hag back said that the bombardier was killed by the 20mm round that made the obvious hole in her breast. *MAFB*

Bugs Bunny munches a carrot on the B-17G, *ROUND TWIP WABBIT*, belonging to the 379th Bomb Group. *MAFB*

*The LUCKY STRIKE,* a B-17F, serial 41-29923, was assigned to the 381st Bomb Group. The name was an interesting play on words. *MAFB*

The B-17F, serial 42-3037, *Windy City AVENGER*, started its career with the 384th Bomb Group and later transferred to the 305th Bomb group. *MAFB*

This lady has her man and isn't letting go. "*Hustlin' Hussy,"* a B-17F, serial 42-30354, was assigned to the 549th Bomb Squadron, 385th Bomb Group. "*Hustlin' Hussy"* was shot down by fighters on a mission to Frankfurt and crashed near Charleroi. *Rukes*

Pictured with the B-17G, *SKY CHIEF* is 2nd Lt. Keiste, of the 385th Bomb Group. *Ethell*

The ground crews stayed very busy keeping all parts of the aircraft in working order, which was no easy task considering the missions they flew. In this shot the camera caught the crew pulling maintenance on the top turret of *SUGAR PUSS*, a B-17F, piloted by Lt. Clarence Christman of the 544th Bomb Squadron, 384th Bomb Group. *Sugar Puss* was shot down by fighters on a mission on 25 July 1943 near Hamburg. *AFM*

*"SACK TIME,"* a B-17F, serial 42-5914, was assigned to the 385th Bomb Group, Eighth Air Force. This photo was taken at Great Ashfield Air Drome in England before the aircraft, piloted by 2nd Lt. Leslie Reichardt, was shot down by fighters. The ill-fated B-17 flew the Schweinfurt/Regensburg mission 17 August 1943. *Lindsey*

There were different variants of dragons and ladies used as artwork on the aircraft from World War II and continuing up to today. The 385th Bomb Group's B-17, serial 42-30836, *Dragon Lady*, showed their lady riding on a dragon. The 551st Bomb Squadron aircraft was damaged by flak and forced to ditch in the channel 13 February 1944 on a mission to Pas'De Calais. *Linsey*

*SCARLETT O'HARA*, a B-17F, was assigned to the 379th Bomb Group. *AFM*

This 385th Bomb Group shark mouth B-17G, serial 42-3547, was named *LATEST RUMOR*. It pictures the girl back home as she spent her spare time on the phone. *Linsey*

*"Tondelayo"* was the name on this B-17F, serial 42-29896, assigned to the 527th Bomb Squadron, 379th Bomb Group. This aircraft was listed as missing in action on 6 September 1943 when it was damaged by fighters and was forced to ditch in the channel while on a mission to Stuttgart. *MAFB*

This B-17 F, serial 42-30096, was one of several aircraft named *Liberty Belle*. It belonged to the 549th Bomb Squadron, 385th Bomb Group. The guns on this fortress have seen action as is indicated by the five swastikas above the fifteen bomb mission markings. The aircraft, piloted by Capt. Robert Smith, had mechanical failure and ditched in the North Sea 30 November 1943. *Ethell*

*Slo-Jo*, a B-17F, serial 42-30168, was piloted by Lt. T.E. Jennings, of the 550th Bomb Squadron, 385th Bomb Group. *Slo-Jo* was damaged by fighters and ditched in the North Sea 11 December 1943. *Ethell*

The first in a line of marching ladies turns to see the soldier who is giving a quick pat in the art on the B-17G, *HIT PARADE Jr.* At the time this photo was taken there had been five German aircraft downed by this ship belonging to the 385th Bomb Group. *Linsey*

This 157 mission B-17 is seen as she sits idled by the end of the war. The 524th Bomb Squadron, 379th Bomb Group's B-17G was among the other retired aircraft at Kingman, Arizona awaiting the smelters when this photo was taken. *Grantham*

*Shoo Shoo Shoo Baby*, a B-17G, serial 42-32076, was assigned to the 401st Bomb Squadron, 91st Bomb Group. This aircraft was damaged by flak 29 May 1944 and crash landed in Switzerland. She was brought back to the states where she was returned to flying condition at the Paul Garber Facility. She was then flown to Wright Field for permanent display. *Rukes*

*Our Gal Sal*, a B-17G, was assigned to the 390th Bomb Group. Note the red bombs used to denote each set of five bomb mission making then easier to count. This aircraft shot down two attacking enemy aircraft. *Ethell*

*V MAIL Gets There Fastest*, a B-17F, was assigned to the 306th Bomb Group. Note the Distinguished Flying Cross painted on the nose. *Ethell*

A cat may have nine lives, but this one seems to need all of them. Sporting a cat on a cloud preparing to drop a bomb is *OH! HARDLUCK*, a B-17G-70-BO, serial 43-37775, assigned to the 96th Bomb Group. Note the red field-dressing on the cat's tail. *Ethell*

*NAUGHTY NANCY*, a B-17G, serial 43-39116, carries a painting of a very scantily clothed reclining female. She is wearing a pair of nylons which was one of the many items which became hard to locate during the war. *Ethell*

*Fools Rush In*, a B-17G, serial 42-31066, was assigned to the 351st Bomb Squadron, 100th Bomb Group. This aircraft was lost 31 December 1944 when it was struck by flak and crashed at Radegast. *Ethell*

*KNOCKOUT DROPPER*, a B-17G, was assigned to the 391st Bomb Squadron, 34th Bomb Group. The aircraft carried many versions of ladies in champagne glasses, this being one. *Ethell*

*"LAY or BUST"*, a B-17G-45-BO, serial 42-97230, proj. #92644-R, was assigned to the 100th Bomb Group. *Ethell*

*BUTCH*, a B-17G, was assigned to the 391st Bomb Squadron, 34th Bomb Group. In addition to the many bomb missions flown by *BUTCH* it also flew five cargo drops. Note the one kill mark on the nose. *Ethell*

*HUMPTY DUMPTY*, a B-17G, was assigned to the 100th Bomb Group. This art was seen on several aircraft with different names accompanying it. Note the colors used for these bomb mission markers. They also used a different color to separate each set of five. *Ethell*

The 5,000th B-17 built was aptly named, *5 GRAND*. The B-17G, serial 43-37716, was covered with the names of all the employees of Boeing before being delivered to the 96th Bomb Group. *Ethell*

*OLE-TIMER*, a B-17G, serial 43-38138, served with both the 34th and 490th Bomb Groups. At the time this photo was taken *OLE-TIMER* was wearing the green nose band of the 391st Bomb Squadron, 34th Bomb Group. *Ethell*

*QUEENIE*, a B-17G, was assigned to the 391st Bomb Squadron, 34th Bomb Group. A three-leaf clover painted on the front of the chin turret matches the green nose band marking the 391st Bomb Squadron's B-17. *Ethell*

This shot was taken at Kingman and shows the B-17G, *Milk Wagon*, after her engines had been removed. *Milk Wagon*, serial 43-37756, had served with the 447th Bomb Group before retirement. Note the white milk bottles used to mark bombing missions have begun to fade in color. *Grantham*

The 91st Bomb Group's B-17F, *Memphis Belle*, serial 41-24485, stands safe at home in the States after being honored as the first B-17 in the 8th Air Force to complete twenty-five combat missions. The odds were very bad for the early missions deep into enemy territory as the fighters didn't have long-range capabilities and had to turn back leaving the bombers unescorted. *Ethell*

This B-17G, complete with its own red head, was possibly assigned to the 4th Bomb Squadron, 34th Bomb Group as it wears the same white nose band. *Ethell*

A Varga girl adorned this B-17 which was also a possible 4th Bomb Squadron, 34th Bomb Group aircraft. *Ethell*

*D-DAY DOLL*, a B-17G, shows the red cowlings of the 710th Bomb Squadron, 447th Bomb Group. This aircraft survived the war only to return to the smelter at Kingman, Arizona. *Ethell*

*COCK O' THE SKY*, a B-17G, was assigned to the 710th Bomb Squadron, 447th Bomb Group. Note the red engine cowlings red lettering, and red mission markers. *Ethell*

*Blonde Bomber II*, a B-17G, was assigned to the 710th Bomb Squadron, 447th Bomb Group. Like *COCK O' THE SKY*, this B-17 also has the red engine cowling, red mission markers and red lettering to match. *Ethell*

Strapped to a bomb on this B-17G is what appears to be a winged pixie in pointed shoes and red pants scoping out the area below for the target of the day. *Ethell*

If your aircraft must have its gremlin why not a female gremlin, *FIFINELLA*, as seen on this B-17F, serial 42-107030, belonging to the 322nd Bomb Squadron, 91st Bomb Group. This was the chosen insignia of the lady pilots who shuttled the aircraft from the states to the awaiting bomber crews. *FIFINELLA* ended her career when she was hit by flak and crashed near Epreville. *Ethell*

*LITTLE Miss Mischief*, a B-17G, serial 42-97880, was assigned to the 91st Bomb Group. It is shown here were it crash landed in Bassingbourn, England, in April of 1945. *Davis*

Sitting in a full champagne glass among the bubbles this female was found on the B-17G, *BLUE CHAMPAGNE*, which was a member of the 385th Bomb Group. *Ethell*

As can be understood, fate was on the minds of all the bomber crews. The crew of this 385th Bomb Group's B-17F, serial 42-3335, named it *FICKLE (finger is understood) of ?. Ethell*

A well armed angel adorns *PREGNANT PORTIA*, a B-17F, serial 42-30263, belonging to the 385th Bomb Group. *Lindsey*

*The Passionate WITCH*, a B-17F, sporting a nude female riding a broom, was assigned to the 385th Bomb Group. *Lindsey*

The noble American eagle is shown swooping down toward the enemy with a bomb in each talon and carrying an armed passenger on this 385th Bomb Group B-17G, serial 42-3544, *Stars and Stripes. Ethell*

*Belle of the Blue* was one of the 358th Bomb Group's B-17s. It was serial number 42-30094. *Lindsey*

*TIGER GIRL* was a shark mouth B-17G, serial 42-3555, belonging to the 388th Bomb Group. Note the five victory crosses. *Ethell*

What evil is Satan working on these two ladies which adorned the B-17G, *HELL'S-BELLES*? The o/d aircraft was assigned to the 388th Bomb Group. *Essig*

This spunky rooster on *COCK O' THE WALK* digs into a falling bomb on the B-17F, serial 42-30800, which was assigned to the 388th Bomb Group. The 388th lost this aircraft 29 February 1944 on a mission to Brunswick when it was hit by flak and crashed at Bienrode. *MAFB*

A less noble member of the eagle family, the vulture on this 388th Bomb Group B-17, serial 42-39845, *Hulcher's VULTURES*, sent a clear message to the enemy. The 563rd Bomb Squadron aircraft was piloted by 2nd Lt. Marquis Fjelsted. On 28 May 1944, while on a mission, it was struck by flak and crashed at Wiera, Germany. *Ethell*

One of the more daring pieces of art was on *MISS CARRY*, a B-17F, serial 42-30325, assigned to the 390th Bomb Group. The use of the word "censored" was quite appropriate for this nose art. *MAFB*

*"LUCKY PARTNERS"*, featuring the singing duet of Jose' Carioca and Donald Duck, was assigned to the 447th Bomb Group, Eighth Air Force. *MAFB*

*HOMESICK ANGEL*, a B-17G, serial 42-31591, started its service with the 401st Bomb Group and later transferred to the 457th. *Lindsey*

This B-17F belonging to the 570th Bomb Squadron, 390th Bomb Group was *CASH & CARRIE*, serial 42-5915. The aircraft, piloted by Lt. A. McGuire Jr., was attacked and shot down by fighters on the Munster raid 10 October 1943. *Ethell*

*GUNG HO*, a B-17G, serial 42-31134, was assigned to the 569th Bomb Squadron, 390th Bomb Group. It joined the number of bombers that were victims of flak damage when it crashed at Nuremburg 10 September 1944. *MAFB*

*COCK O'THE SKY*, was a bare-metal B-17G sporting a fighting cock on the side. It was one of the B-17s belonging to the 447th Bomb Group. *MAFB*

The 447th Bomb Group's B-17G, *L'IL EIGHT BALL*, shows numerous bomb mission markers on the lower portion of nose. *Ethell*

The missions of this B-17G, serial 43-37756, were marked with milk bottles. *Milk Wagon* was assigned to the 447th Bomb Group. *MAFB*

*Calamity Jane* was one of the B-17s returned to the states, from the 13th Air Depot Group on New Caledonia in the Pacific, for use as a training aircraft after completing its combat assignment. This photo of *Calamity Jane* was taken at Amarillo Army Air Field. *Snodgrass*

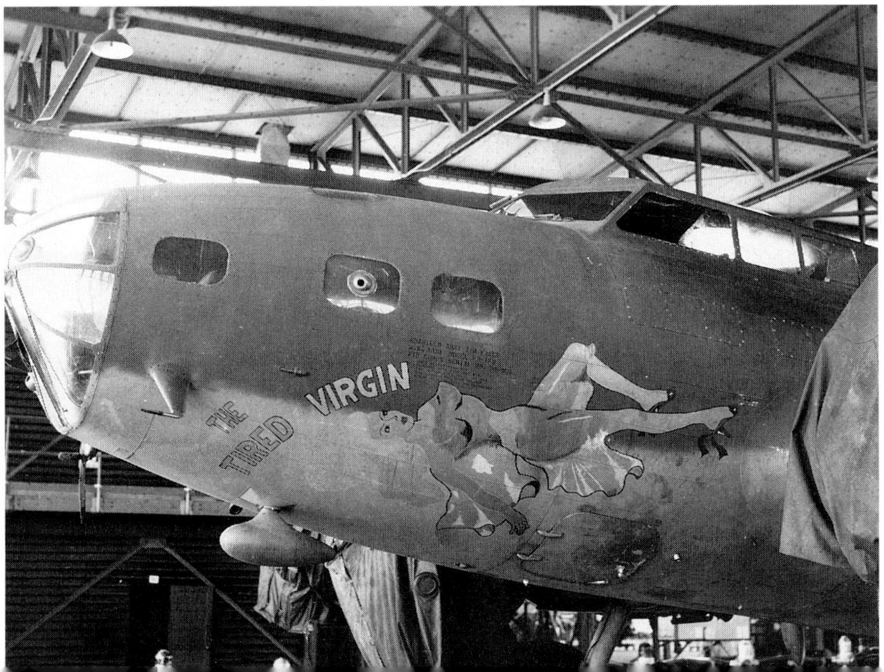


This U.S. Army Model RB-17B, Air Corps serial 39-006, *THE TIRED VIRGIN*, was stationed at Amarillo Army Airfield. You might note the artwork on the aircraft in the states was not as risque as it was one the aircraft overseas. This painting appears to have been retouched to provide a less shocking effect to the viewer. *Snodgrass*

The well armed Flying Fortress not only had nose art but in some cases also had names for the twin .50s in the tail gunner position. Three examples being *The WOOFLEDUST TWINS*, *LOUISIANA DIPLOMATS*, and *RAT POISON*, which shows one victory. *Snodgrass*

The bomber crews could breath easier while the fighters were providing them an umbrella of protection. So to honor their "Little Friends," the crew of this 452nd Bomb Group named this B-17G, *Our Buddy. Ethell*

This demon from hell was on the nose of the B-17F named *the RED GREMLIN. MAFB*

This B-17F, sporting a reclining female, was named *STOP AND STARE* by its crew members. *MAFB*

*Miss WINDY City*, a B-17G, features a female caught in a very strong wind. *MAFB*

*TET TITTOTE*, a B-17, carries its name in a Chinese motif. In contrast, however, on the nose it carries a painting of the American Bald Eagle. *Ethell*

144
NACA
FERTILE MYRTLE
P2B-1S
NAVY
ART STAGG
TOM CONGDON
ROGER BROWER
VINCENT X GOLDEN
PHIL MACHADO
TONY MACHADO
KRON TV KTVU TV
EXPERIMENTAL

## Chapter Four

# Super Fortress

The B-29 is best known as the first aircraft to deliver an atomic bomb. This huge aircraft with its crew of eleven, met and exceeded most records set by aircraft of its day. It not only saw combat in World War II but also in the Korean conflict. It's service after the war years placed the B-29 as the most versatile bomber of its day. The Superfortress was used to drop the test ships, and rocket power was also introduced with the aid of the Superfortress. "Glamorous Glennis", Bell's X-1, dropped from the underbelly of a specially modified B-29, was the first aircraft to break the sound barrier. Parasite fighters were carried by the B-29, also bearing significant results in research.

Many men flew in the B-29 and have literally thousands of stories to tell. The B-29 was the first aircraft to be sectionally pressurized. It could fly higher and further than any other heavy bomber of World War II.

A total of 3,970 B-29's were built. Among its assignments the B-29 flew the Hump into China, and because of its range, the longest of any World War II era bomber, was able to fly further and deeper into the heart of Japan. Its dropping of the first atomic bomb helped shorten the war and save American, as well as Japanese, lives. The first mission against Japan flown by a Superfortress, came on the June 5, 1944, when a mission was launched to bomb Bangkok, Thailand. A little over a week later the B-29s again massed to bomb Yawata, Japan.

Called "LeMay's Bomber", the B-29 has etched a legacy into the hearts of the men who flew and serviced her, and has become a legend in aviation history.

*FERTILE MYRTLE*, a U.S.Navy P-2B-15, assigned to duties with NACA, what later became NASA. This aircraft was used as is visible from the scoreboard as a drop ship or mother ship to drop the Naval flight test Skyrocket. This photo depicts 144 drops. Quite an impressive record for the Superfort as well as the test program. *Veronico*

The B-29 *There'll Always Be a Christmas* carried a well known Varga girl who was accompanied by this verse written by Phil Stack. *Campbell*

*DOUBLE EXPOSURE*, an F-13A-55-BW, serial 42-24877, piloted by Capt. Thomas C. Kendall's was a member of the 3rd Photo Recon Squadron. *Campbell*

*Poison Ivy*, an F-13, serial 42-24585, commanded by Colonel Patrick McCarthy, flew with the 3rd Photo Recon Squadron. The art was named for the aircraft's crew chief, Flip Ivey. *Scheffe*

*DESTINY'S "TOT"*, a B-29, serial 42-65284, flew as lead ship for the 9th Bomb Group. *Stauffer*

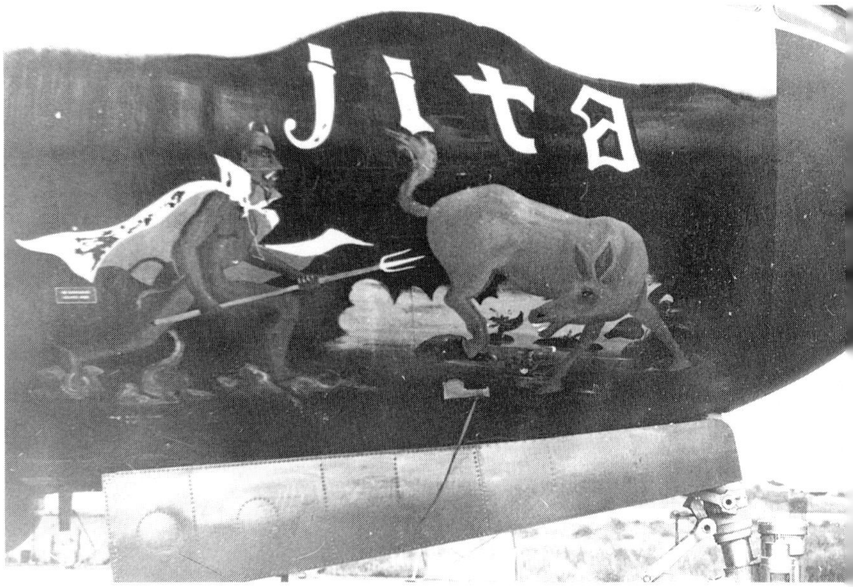


*jita*, a B-29, assigned to the 93rd Bomb Squadron, 19th Bomb Group. The name was the short version of *jab in the ass. AFM*

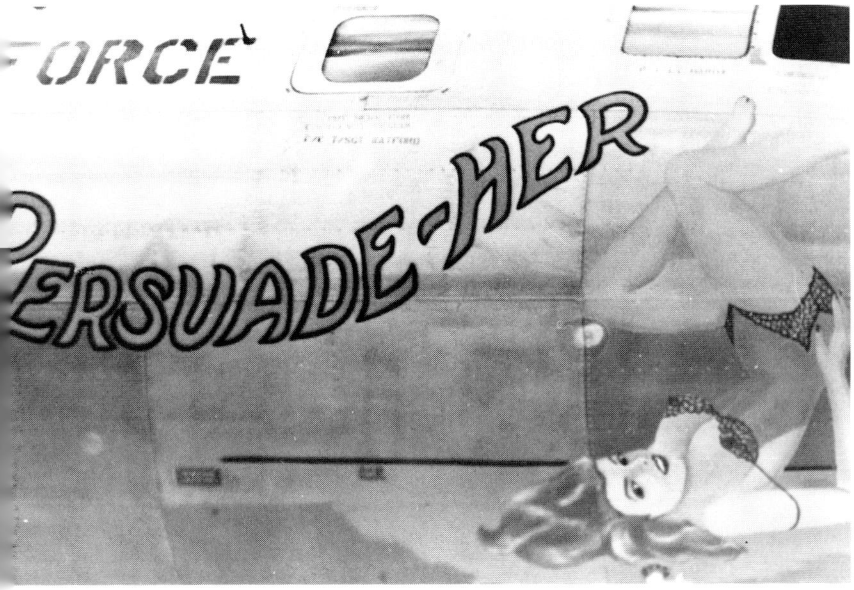


*PERSUADE-HER*, a B-29 which was assigned to the 28th Bomb Squadron, 19th Bomb Group. *AFM*

The B-29, *Dead Jug*, assigned to the 19th Bomb Group, portrayed the easy life, with the hillbilly and his moonshine jug, bare-foot, resting under a tree. The term "dead jug" was intended as a derogatory reference to the problems with the R-3350 engine. *AFM*

*OUR GAL*, a B-29, 19th Bomb Group which was stationed at North Field, Guam. *AFM*

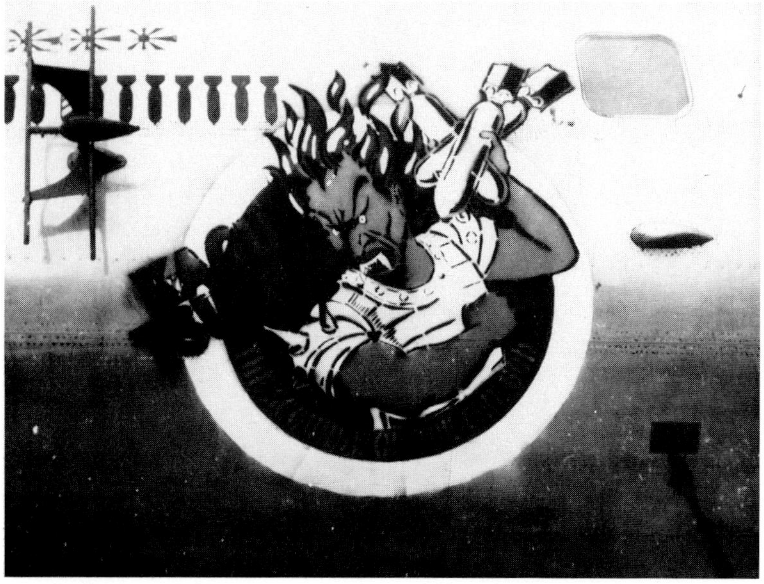

This insignia was carried by the 40th Bomb Group's B-29s. *KAGU TSUCHI/ THE SCOURGE OF THE FIRE GOD* was the accompanying slogan. This B-29 boasts of shooting down three Japanese aircraft. *Stauffer*

*Miss Spokane*, a B-29, serial 44-27332, flew with the 92nd Bomb Group and was passed on to the 98th Bomb Group. *Campbell*

*Winnie II*, a B-29, serial 44-69963, was also assigned to the 444th Bomb Group based at West Field, Tinian during 1945. *Campbell*

*Patience Reward*, a B-29, serial 44-70131, was assigned to the 444th Bomb Group on Tinian. This aircraft shows twenty-one bomb missions including one with a heart designating a mission into the heart of Japan. *Campbell*

*HO-HUM/"LET'S DO IT AGAIN!"* was a B-29, serial 44-70123, which was assigned to the 444th Bomb Group. Applied later to the nose was the 676th Bomb Squadron insignia featuring the Disney character *The Reluctant Dragon*. *Stauffer*

The B-29, *Miss Patches*, was assigned to the 444th Bomb Group. *Campbell*

*MUSN' Touch it*, a B-29, was assigned to the 444th Bomb Group. *Stauffer*

*VICTORY GIRL*, a B-29, serial 42-24731, was a member of the 444th Bomb Group. *Stauffer*

*SLEEPY TIME GAL*, a B-29, was a member of the 444th Bomb Group. *Stauffer*

*Miss You*, a B-29, served with the 444th Bomb Group on Tinian. *Stauffer*

*Sky Queen*, a B-29, was assigned to the 444th Bomb Group. *Stauffer*

*"Celestial Queen"* was among the B-29s belonging to the 444th Bomb Group. *Stauffer*

*SHAFT ABSORBER*, a B-29, serial 44-48068, was assigned to the 444th Bomb Group. *Stauffer*

*Lady in Waiting*, was the same aircraft as *SHAFT ABSORBER*. They shared the art but had different names. *Campbell*

*SURE THING*, a B-29, serial 42-63557, was assigned to the 444th Bomb Group. *Campbell*

*SLAVE GIRL*, a B-29, serial 44-27307, flew with the 444th Bomb Group. *Campbell*

*DREAM GIRL*, a B-29-21-BA, serial 42-63480, was assigned to the 444th Bomb Group. *Stauffer*

The stork on *AIRBORN* carries a welcome gift, which was not true of the B-29, serial 42-65268, belonging to the 444th Bomb Group. It shows thirty-seven plus bomb missions, three trips over the hump and one photo recon mission. *Campbell*

*SKY BLUES*, a B-29, was among the aircraft of the 444th Bomb Group. *Campbell*

The 444th Bomb Group B-29, *HIGH and MIGHTY*, shows a female holding high the 20th Air Force insignia. On the nose was the insignia of the 677th Bomb Squadron, a western style Uncle Sam in sheriff's garb. *Stauffer*

A scantly clothed maiden reaches up to touch a flak burst on the B-29, serial 44-70129, *FLAK MAID*, which served with the 444th Bomb Group based on Tinian. *Stauffer*

The sort of censoring of the art seen on the 444th Bomb Group B-29, *Lady Frances*, usually took place when the aircraft was homeward bound or after it had returned to the states. Note the flight of six B-29s painted just below the bomb missions. *Stauffer*

The art work titled *Naval Manuvers* was chosen for the B-29, *"This is IT."* This aircraft, serial 42-6321, was among the B-29s belonging to the 444th Bomb Group stationed on Tinian. *Stauffer*

*HELLON WINGS*, a 444th Bomb Group B-29, sports a winged female. This aircraft has been over the hump five times and gone on twenty-two bomb missions including two into the heart of Japan. *Campbell*

*Lady be good*, a B-29, serial 42-65227, was assigned to the 792nd Bomb Squadron, 486th Bomb Group. This aircraft shows five trips over the hump and one Japanese plane shot down. *Stauffer*

*Memphis Maid*, a B-29, serial 44-70120, was assigned to the 444th Bomb Group. *Campbell*

*up n' comin'*, a B-29, was assigned to the 444th Bomb Group. *Campbell*

This lovely lady , possibly named *LOUISE*, was on a 444th Bomb Group B-29. *Campbell*

*"SO TIRED"/SEVEN TO SEVEN*, was an RB-29A-45-BO, serial 44-61727, which served with the 91st Recon Squadron in Japan during the winter of 1950. The photo was taken by Lt.Col. Mike Moffett. *Taylor*

*NO SWEAT*, a B-29, served with the 28th Bomb Squadron, 19th Bomb Group. After crash-landing in Korea, the aircraft was finished off by a P-51 which suffered engine failure on take-off and crashed into her. *Watts*

*AH SOOO. . . .*, an RB-29A-50-BO, serial 44-61817, was piloted by 1st Lt. David Newcomb of the 91st Recon Squadron. The photo was taken by Lt.Col. Mike Moffett in Japan during the winter of 1950. *Taylor*

*HONEYBUCKET HONSHOS*, an RB-29A-55-BO, serial 44-61929, was assigned to the 91st Recon Squadron. The honey buckets carried caricatures of the crew members. This picture was taken during the winter of 1950 in Japan. *Moffett*

This is another of the photos taken, of the aircraft assigned to the 91st Recon Squadron, by Lt.Col. Mike Moffett. This photo was taken of the RB-29A-55-BO, serial 44-61951, *OUR L'LASS* during the winter of 1950, while it was stationed at Yokota, Japan. *Taylor*

*WHERE NEXT?* an RB-29A, serial 44-613992, was piloted by Lt. J.E. Ernest, of the 91st Recon Squadron. The photo was taken during the winter of 1950 at Yokota, Japan. *Taylor*

*MOON'S MOONBEAM*, an RB-29A, serial 44-611815, was assigned to the 91st Recon Squadron. This photo was taken at Yokota, Japan during the winter of 1950, by Lt.Col. Mike Moffett. *Taylor*

This RB-29A, serial 44-611727, sports the insignia of the 91st Recon Squadron. The art was that of Lt.Col. Mike Moffett Photo Officer for the 91st. *Moffett*

The cartoon character, Woody Woodpecker, is seen on this B-29, *FIRE BALL*. The complete art shows a lady being chased by Woody. *Conners via James*

*THE FRY'IN PAN*, a B-29, was named for its pilot whose last name was Fry. *Conners via James*

*SHADY LADY*, a B-29, serial 4265357, was assigned to the 98th Bomb Group. The artist was effective in his use of the silhouette in this more unique piece of nose art. *Shady Lady* created a provocative mood, yet remained a discreet piece of art of which no commander could disapprove. *Conners via James*

*"JACK POT,"* a B-29, serial 42-24797, was assigned to the 505th Bomb Group. *"JACK POT"* crashed into the ocean after loosing two of its engines on a mission in March of 45. After drifting two days the crew was spotted by a Super Dumbo and rescued by the destroyer *Gatling*. *Tinker AFB*

"JACK POT"

*GOIN' JESSIE*, a B-29, serial 42-24856, was assigned to the 9th Bomb Group.
*Stauffer*

*AL-ASK-HER!*, a B-29, serial 45-21760, served in the Korean War. Many of the missions were flown at night. This paint scheme made it difficult for the North Koreans below to see the aircraft in the night skies. *Witt*

*WACHIN JO*, a B-29, serial 45-21717, was another of the aircraft which served during the Korean War. Keeping with the bear in the art, all of the missions markers on the nose were small white bears. *Witt*

There were several different B-29s with the name, *OUR GAL*. This one appeared on serial 42-24484 which fought during World War II. *Campbell*

*CITY OF QUAKER-CITY* was flown by Lt. Lester Gilbert. This B-29 flew 23 missions in the 314th Bomb Wing before being put out to pasture. *Davis*

*SENTIMENTAL JOURNEY* was the name on the Left side of the B-29 also known as *CITY OF QUAKER-CITY*. *Davis*

*THE SPEARHEAD*, a B-29, serial 44-69975, was assigned to the 9th Bomb Group. Formally known as *Man O'War II*. The art was of the very famous photo which commemorated the flag raising on Iwo Jima by the 5th Marine Division. *Stauffer*

*Eddie ALLEN*, a B-29, serial 42-24579, was assigned to the 40th Bomb Group. This aircraft was named for the Boeing chief test pilot who made the first flight in the XB-29 and was killed during a test flight of the 2nd prototype. During her career *Eddie ALLEN* participated in the bombing raids of seven countries. After suffering serious damage on May 25, 1945 during a mission to Tokyo, she was sent to salvage where she was used as a ditching trainer, as pictured here. *Stauffer*

This lovely lady was seen on both sides of the 444th Bomb Group B-29, serial 42-24863, *LUCKY LADY*. *Stauffer*

This version of the Disney character, *THUMPER*, was seen on the B-29 belonging to the 504th Bomb Group stationed on Tinian. *Campbell*

Although it is amongst these B-29s, the 312th Bomb Group, *The Lady is Fresh* was a B-32 Dominator. It was one of only three of its kind to see combat. The serial on this aircraft was 42-108529. *Acord*

*PHONY EXPRESS*, a B-29, serial 42-24801, flew with the 462nd Bomb Group stationed at West Field, Tinian. *Witt*

*BEDROOM EYES*, a B-29, serial 42-24610, was assigned to the 489th Bomb Group with which she successfully flew sixty-one missions. *Stauffer*

*Miss HAP*, a B-29, serial 42-24774, was assigned to the 497th Bomb Group, Isley Field, Saipan. The aircraft was named in honor of "Hap" Arnold. *Campbell*

*SKY-SCRAPPER*, is an accurate name for this flying eagle wearing boxing gloves and carrying lighting in his mouth. This B-29, serial 42-63463, was assigned to the 497th Bomb Group. *Campbell*

*Jokers Wild*, a B-29, serial 42-24626, was assigned to the 497th Bomb Group, Saipan. *Campbell*

*Our Baby*, a B-29, serial 42-24597, belonged to the 497th Bomb Group. *Campbell*

*Torchy*, a B-29, serial 42-24646, was assigned to the 498th Bomb Group, Isley Field, Saipan. *Torchy* ended her career after an accident 13 April 1944. *Campbell*

*SATAN'S LADY*, a B-29, serial 42-24779, was assigned to the 504th Bomb Group, 313th Wing. On returning from a mission from Kobe it was discovered that flak had punched close to two hundred holes in the aircraft. *Stauffer*

*TOKYO TWISTER*, a B-29, serial 42-24682, flew with the 499th Bomb Group which was stationed at Isley Field, Saipan. *Campbell*

*HONORABLE TNT WAGON*, a B-29, serial was among the aircraft of the 505th Bomb group based at North Field, Tinian. *Stauffer*

*INDIAN MAID*, a B-29, serial 42-24806, assigned to the 504th Bomb Group, is being painted with the Seabees unit insignia to commemorate the construction efforts of the Seabees in building the landing strips used by the B-29s. *Ethell*

*LOADED Dice*, a B-29, was assigned to the 498th Bomb Group. *AFM*

*ENOLA GAY*, better known for her ominous reputation of being the first aircraft to deliver an atomic bomb, she is seen here in a state of repose at ALC/Tinker AFB. The *Enola Gay* underwent modification at Tinker Field and was later delivered to the Paul Garber facility at Suitland, Maryland. Pictured under the cockpit we see the little Fat Man mission markers for missions flown during the Bikini Island bomb tests of the late 1940s and early 1950s, Operation Cross Roads. *Tinker AFB, Office of History*

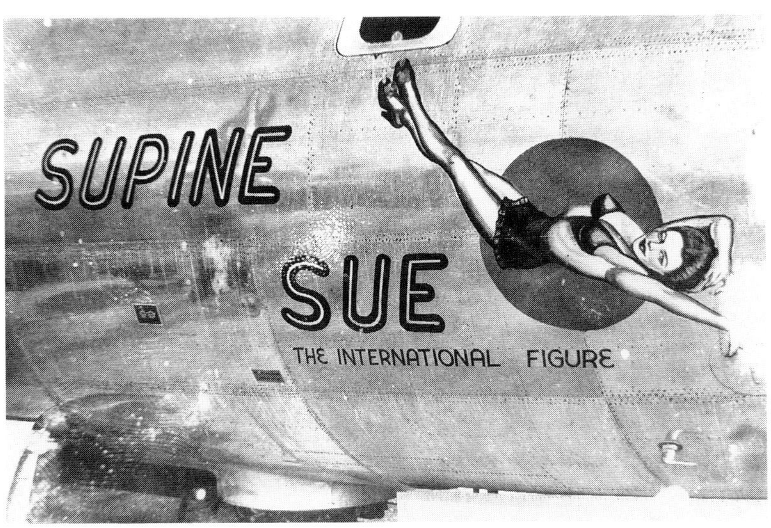


*SUPINE SUE*, a B-29, serial 42-24653, was assigned to the 42nd Bomb Squadron, 500th Bomb Group which was also stationed at Isley Field, Saipan. This aircraft flew forty plus missions. *Ethell*

*PASSION WAGON*, a B-29, serial 42-63524, was assigned to the 498th Bomb Group. *Campbell*

*The Lodger BY GOD*, a P-38, was piloted by Lt. Roger Palmer, who flew with the 432nd "Clover Leaf" Squadron, 475th Fighter Group know as "Satan's Angels." This aircraft owes its name to the following poem: *There was an old maid from Cape Cod, who thought all good things came from God. But was not the Almighty, who lifted her nightie, it was Rodger-the-Lodger by God!* Pictured with *The Lodger* is Lt. R.B. Maxwell. The photo was taken at Kimpo Airfield, during the winter of 45/46 during the occupation of liberated country. This aircraft was among 150 low-time P-38Ls destroyed by a bulldozer in Korea in 1949 to keep them out of the hands of the Koreans. *Maxwell*

## Chapter Five

# Little Friends & Light Weights

Although they were referred to by bomber crews as "little friends," the fighters did a very big job – only their size was referred to in this nickname. Fighter aircraft played both offensive and defensive roles. Pilots of these small agile aircraft were on continuous standby alert. At any time they could be called on to intercept and terminate the activities of attacking Axis fighter or bomber aircraft. At the blare of the siren, the club house and barracks would empty and the sky would soon be full of scrambling aircraft.

Fighters were the best available form of defense against attacking bombers. They could target the enemy more precisely than "ack-ack" guns and of course were more mobile. The main objective of the fighter aircraft was to drop the enemy bombers before they could reach their targets and drop their loads of destruction – the fighters performed this job very well.

Many of the fighters took on still another role – fighter-bombers. For tactical bombing the small aircraft was ideal. Since it carried a smaller bombload than larger bombers, and being smaller and faster it could search out enemy movement and hamper it quite effectively. Fighter-bomber pilots quickly realized they could also use their guns to inflict damage to enemy ground forces and movements. Strafing runs were soon initiated on "targets of opportunity" on the return to base following bombing runs. This proved to be another strong point of the fighters.

These duties were very important to the war cause, though they were not what earned the fighters the name, "little friend." You have only to look at the statistics of losses incurred by bomber crews on runs made without the umbrella of protection provided by the fighters to appreciate the name. A group of bombers made a very good target for enemy fighters. The bomber group had a mission plan and route, and they followed that plan as a group until the job was finished and they returned home, or until it became impossible to do either. Early bomb missions were escorted as far into enemy territory as possible by the short-range fighters but as the fighters peeled off to return to base the odds of survival for the bombers dropped sharply. With the advent of long-range fighters the bombers stood a much better chance of succeeding in their missions, and thus bomber crews also had better odds of surviving. It was this relationship that prompted the name, "little friends."

The various fighter aircraft went through many modifications in war time. Their size and speed made it an excellent choice for many jobs – one being photo missions. In the case of the Lockheed P-38, the guns were removed from the nose and replaced with cameras. Some were "souped-up" and polished to get the added speed needed to evade, which is essential for an unarmed aircraft in enemy territory. Pilots of recon modified North-American P-51s were more fortunate, in that the placement of cameras did not eliminate their guns. Photo recon aircraft flew alone, and for the most part unarmed, and had to fly "on the deck" to get the detailed photos that were needed. Although they were not attack aircraft they flew their missions bravely and many lost their lives.

As you see, the fighter, and its counterpart, the photo recon aircraft, although small, did indeed play a very large role in the fight against tyranny.

# P-38's

*The SOONER The Better*, a P-38, was piloted by Cpt. R.G. Furnbull of the 35th Fighter Squadron, 8th Fighter Group. It was crewed by c/c C.C. Bush and his assistant "Doc" White. The photo was taken in 1945. *Shaw*

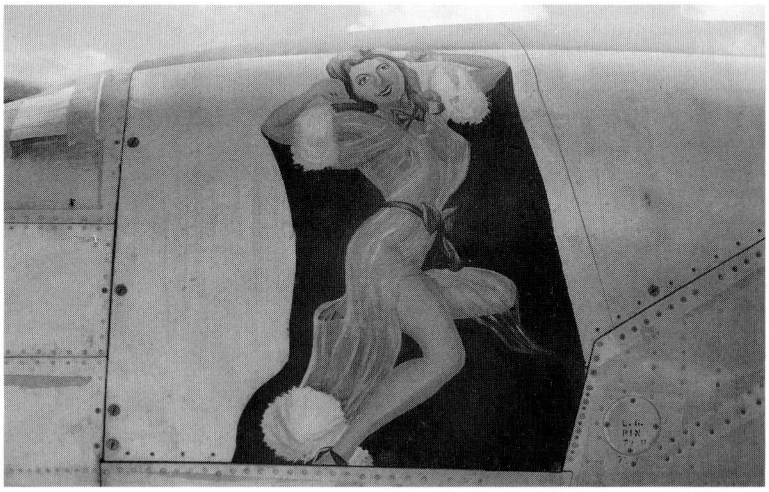

Nameless now, this P-38 had once carried the name *BABY GIRL*. After it was transferred to Cpt. Robert L. Maynard of the 36th Fighter Squadron, 8th Fighter Group, he had the name removed but chose to keep the art. Photo taken on Ie Shima in 1945. *Shaw*

*Vagrant Virgin*, a P-38, was piloted by Lt. L.V. Bellusci of the 36th Fighter Squadron, 8th Fighter Group. The art on the P-38s of the 8th Fighter Group was that of Sgt. Yale Saffro in the 70th Fighter Squadron who was a Disney artist. *Shaw*

Three pillows, possibly red, white, and blue, hold the young lady which adorned Lt. S.L. King's P-38. Lt. King was assigned to the 8th Fighter Group in 1945 when this photo was taken by Sgt. John Shaw of the 4th Photo Charting Squadron. *Shaw*

*Wicked Woman*, a P-38, was piloted by Lt. William C. Meadows of the 36th Fighter Squadron, 8th Fighter Group. On the engine cowl is the insignia of the 36th Squadron, the "flying fiend." *Shaw*

*Jean Creamer*, a P-38 of the 80th Fighter Squadron, 8th Fighter Group. This photo was taken after the new paint, blue over white, was applied. *Shaw*

*G.I. Miss U*, a P-38 belonging to the 35th Fighter Squadron, 8th Fighter Group. Note the black panther inside an oval. This was the insignia of the 35th Fighter Squadron. *Shaw*

*Obscene Corrine*, a P-38, was assigned to the 36th Fighter Squadron, 8th Fighter Group. *Shaw*

*Maximum Goose*, a P-38, was assigned to the 49th Fighter Squadron 14th Fighter Group stationed in North Africa. The pilot of this aircraft was Lt. William J. Hoelle. He is personally credited with downing a Junkers 88, and a Messerschmitt Bf-109G. He was also credited with a probable Bf-109 on December 2, 1942. *Huff*

Nose art found its way onto almost every visible part of the aircraft. We see here a nose wheel cover from a P-38 of the 49th Fighter Squadron, 14th Fighter Group. This ship was assigned to Lt. Al Wolfmuel, thus giving an explanation to the wolf in the art work. *Huff*

This P-38 of the 49th Fighter Squadron, 14th Fighter Group pictures an obvious Texan in a ten gallon hat and chaps, all the gear needed to rule the skies, riding an aircraft as a real Texan would his favorite mount. *Huff*

This P-38 assigned to the 49th Fighter Squadron was flown by Lt. James E. Butler. His name is well represented by the painting on the boom of a gentleman butler. Lt. Butler is credited with shooting down four enemy transport aircraft on 24 Nov 1942 in the vicinity of Gabes, N. Africa. Butler was shot down over N. Africa and lost. *Huff*

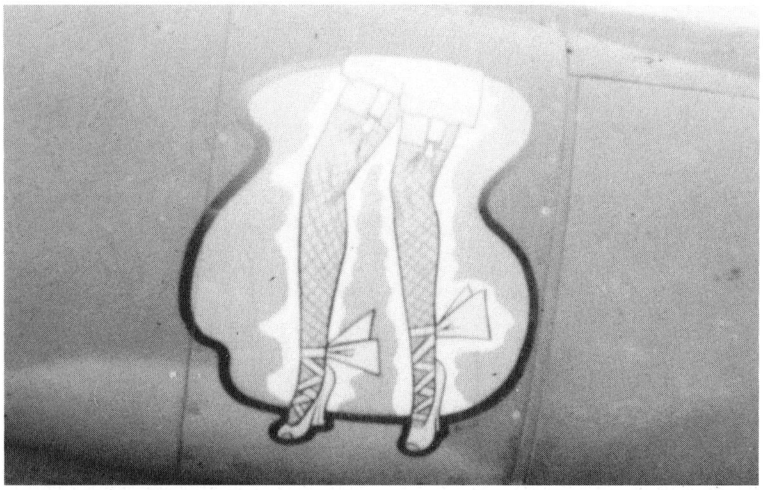

The pretty legs on this P-38 boom art belonged to Lt. Robert N. "Leggs" Carlton of the 49th Fighter Squadron, 14th Fighter Group N. Africa. Carlton is credited with shooting down two German aircraft, a Bf 109 and a Ju-52. *Huff*

The boom art, prevalent to the 49th Fighter Squadron 14th Fighter Group, seemed very common during the early desert war in N.Africa. Seen here we have twin lightning bolts, perhaps significant of the twin tailed P-38 "Lightning" on which this art work rests. *Huff*

Walt Disney was one of the most revered sources for art work on aircraft during the Second World War. We see here Donald Duck in his costume of a devil, complete with forked tail. The 49th Fighter Squadron unofficial patch and slogan was "The Forktailed Devils" coined originally by the German pilots who had to fight against the P-38. *Huff*

*KISKA KICKER/CASANOVA KID*, a P-38G-15-LO, serial 43-2320, was piloted by Lt. Marshall Hyde of the 54th Fighter Squadron, 343rd Fighter Group. Pictured are Lt. John Geddes and Dr. Walter Stager, the flight surgeon. Amchitka Alevtions 1943. *Campbell Archives*

*SCRAPIRON IV*, a P-38, was assigned to the 367th Fighter Group.

*Sweet SUE (alias BO-PEEP)* was one of the P-38s which fought against the Luftwaffe. At the time this photo was taken this aircraft had participated in sixteen bombing missions and shot down two german fighters.

# P-40's

*Above and left*: The 79th Fighter Group's P-40s were all marked with the squadron insignia on the nose leaving little room for art, but Sgt. Joe Pumphrey found the rudder provided the space needed and painted many young ladies on them. They were all originally painted totally nude, but the commander ordered the men to put some clothing on the ladies. Pumphrey added as little as possible as these four photos show. *Sanders*

*"Sand Fly-4th"*, a P-40, is marked with the "flying skull" of the 85th Fighter Squadron, 79th Fighter Group. *Sanders*

This 57th Fighter Group P-40 named *UNCA BUD IV* sits in the North African sand as the pilot poses with his newly painted swastika. It was these surroundings that caused the North Africa fighter groups to think of themselves as *Dessert Rats*. They ate, drank and slept sand. But even in those conditions there was time for celebration as seen on this pilot's face. The art on the tail of the P-40, *UNCA BUD*, was really unique and seemed to cause some embarrassment. It carried a more humorous message for it viewers. *Sanders*

# P-47's

The P-47D, *ANGEL EYES*, serial 42-26272, flew with the 361st Fighter Squadron, 356th Fighter Group in the ETO. *Heitzel*

Fighters, although they made small targets, feared the damage done by flak, as they couldn't handle it as well as the heavies, who often returned with large tears in their skins. This concern is reflected in the P-47, *FLAK SACK*, assigned to 4th Fighter Group. *Cullember*

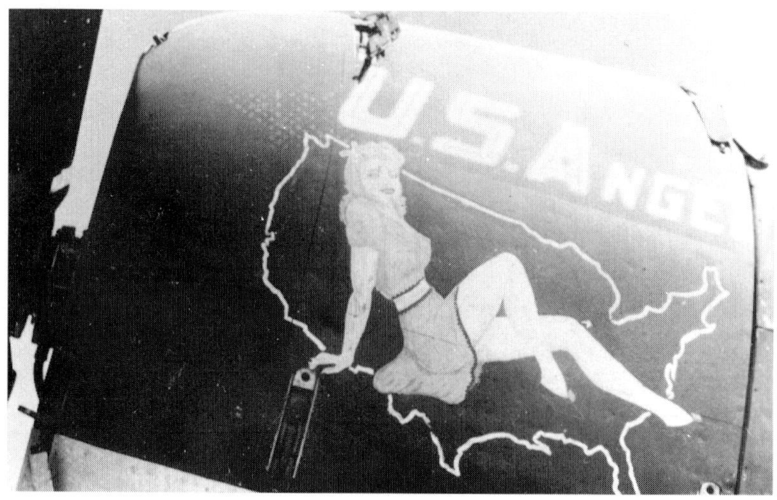

Sitting in an outline of the states, a smiling female adorned this Pacific P-47 named *U.S. ANGEL. MAFB*

This pretty blonde was on a P-47 assigned to the 79th Fighter Group. The 79th used their P-47s in the traditional role, both as an escort and on low level strafing/bombing missions. *Sanders*

*Classy Chassy* was another of the P-47s of the 79th Fighter Group. *Sanders*

*Hula Girl*, a P-47, was assigned to the 79th Fighter Group. Note the wavy letters denoting the dancing movement in this art. *Sanders*

*Evie*, a P-47, assigned to the 405th Fighter Group sports a young lady which looks as if she just came off a valentine card. *Davis*

*TIPSY*, a P-47, was assigned to the 405th Fighter Group. *Davis*

*Touch of Texas*, a P-47, was assigned to the 405th Fighter Group. The pilot of this aircraft used one of the more feared and respected predators of his home state to adorn his P-47. *Davis*

*BAMA*, a P-47, assigned to the 405th Fighter Group. This elephant is a good comparison to the P-47. As the elephant stands alone in size in the jungle, the massive size and fire power of the P-47 set it apart from the rest of the fighters. *Davis*

A fierce looking funnel adorned the P-47, *Kansas Tornado II*, which belonged to the 405th Fighter Group. *Davis*

*YANKEE Tarheel*, a P-47, was assigned to the 405th Fighter Group. *Davis*

Comic book characters were chosen as nose art favorites by many pilots. A Dogpatch character, *Daisy Mae*, was painted on this natural metal P-47 which served with the 56th Fighter Group, Zemke's Wolfpack, in the ETO. *Ethell*

*RAID HOT MAMA*, a P-47, was assigned to the 405th Fighter Group. *Davis*

*Parson's Wife*, was another of the a P-47s assigned to the 405th Fighter Group. *Davis*

*THE BUG*, a P-47, was assigned to the 405th Fighter Group. *Davis*

*TING A LING*, a P-47, was assigned to the 405th Fighter Group. *Davis*

Flying the skull and cross bones is the P-47, *THE GRIM REAPER* which was assigned to the 405th Fighter Group. *Davis*

*Luger Luggin Lulu*, a P-47, was assigned to the 405th Fighter Group. *Davis*

*BALLS OUT*, a name seen on several World War II aircraft, accompanied the charging bull on this P-47D, serial 44-33813, assigned to the 405th Fighter Group. *Davis*

# P-51's

The P-51D, serial 44-72322, named *NANCY LEE*, was piloted by Lt. R.L. Baggett who was assigned to the 402nd Fighter Squadron, 370th Fighter Group, 9th Air Force. *Jones*

This photo of the aircraft belonging to the 363rd Fighter Squadron, 357th Fighter Group, was taken in Leiston, England in 1944 prior to it being shot down with Capt. Hershel Pasco at the controls. The pilot of the P-51D-5-NA, *DESERT RAT*, was intimately acquainted with the effect of the desert wind as portrayed by the frayed lettering on the name of his plane. Note the palm trees and tent that complete the desert motif.

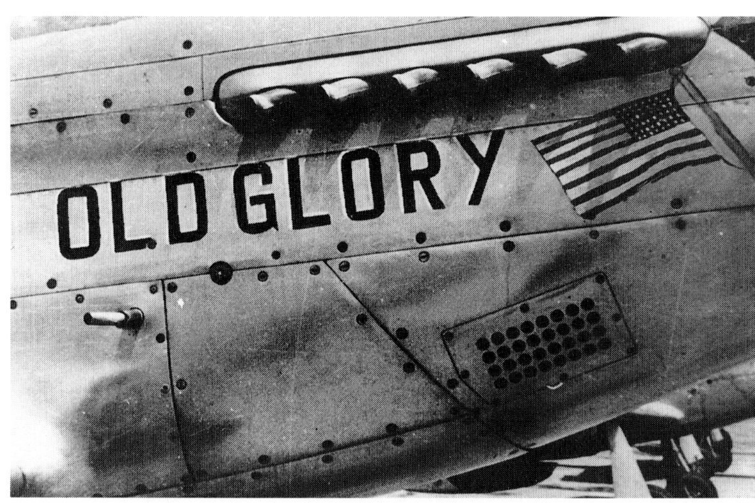


Never is the flag more important to us than when it is threatened. No one knows its value more than the person who is risking all to protect it. It seems very fitting that this P-51, named *OLD GLORY*, shows the flag still flying high and proud.

Taxiing down the runway is the P-51, *DEUCES WILD*. It shows the blue band with white stars on the cowling identifying it as one of the aircraft of the 356th Fighter Squadron, 354th Fighter Group. *Watts*

# Liaison

Standing beside *TWIN MAGS*, a liaison aircraft seen here on Guam in 1945, is U.S. Navy's Able Seamen, Donald Theime. *Theime*

*Miss. Bee Havin* was another of the liaison aircraft used as a squadron hack. This one belonged to the 458th Bomb Group. *Heitzel*

# Captured

This German bomber, a Ju-88A, was captured by the 86th Fighter Squadron and now wears their squadron insignia. While with the 86th it served as a squadron hack. After the war it was among the aircraft returned to the states and is currently on display at the museum at Wright Field. *Pape*

This captured Macchi MC.200, Italian fighter, pictured in North Africa, carries the insignia of the 86th Fighter Squadron, The Comanches. *Hoffman via Sanders*

*Rosie*, an F-86, served with the 4th Fighter Interceptor Wing in Kimpo, Korea 1953. *James*

This F-5 has art on the engine cowling. *Soderlund via Copeland*

This full shot of *MODIFICATION*, #331, taken at Mt. Farm, England in July 1944, shows the invasion strips on the boom. Paul Hartung was the crew chief and Herman Schneeman was asst. crew chief of this F-5. *Holmes*

This shark mouth F-5E-2-LO, serial 43-28619, was piloted by Lt. Rickey who was a member of the 7th Photo Recon Squadron. His crew was made up of c/c Orin Young and asst. c/c Roger Carnes. Lt. Rickey was reported missing in action 26 November 1944. In February 1945 he was reported to be a prisoner of war. He later returned to England in May 1945. *Holmes*

*HE LIL' DAWG!*, a P-38J, showed the red spinners of the 21st Fighter Group. These P-38s were used by the 318th Fighter Group on Saipan. *Stauffer*

*Peg O' My Heart*, a P-38, was assigned to the 21st Fighter Group. This photo was taken on Tinian. *Stauffer*

*MODIFICATION*, an F-5C-1-LO, serial 42-67331, was piloted by Lt. McDonald who was assigned to the 7th Photo Recon Squadron, 2nd Photo Recon Group, Eighth Air Force. The photo depicts the removing of the guns of the P-38 and the addition of the cameras, which changed the fighter into a recon aircraft with the designation F-5. The recon aircraft faced great danger and faced it unarmed. This aircraft was lost 24 September 1944 and its pilot, Lt. McDonald was listed as missing in action. *Holmes*

*NO NOTHING*, an F-5E-2-LO, serial 43-29000, was piloted by Lt. Stamey, of the 7th Photo Recon Squadron. Lt. Stamey's crew was c/c Vernon Carlson and asst. c/c Charles Modlin. Lt. Stamey was reported missing in action 19 September 1944. Photo taken at Mt. Farm, 1944. Note the reflection of the Marston matting (Mars matting) in the nose lower camera bays. *Holmes*

*"8" BALL*, an F-5E-2-LO, serial 42-28999, was piloted by Capt. Florine, of the 7th Photo Recon Squadron. The art on this aircraft reflects the alliance of the pilot and crew of an aircraft by depicting all three together behind the 8 Ball. Florine's c/c was Sam Hoagland. His asst. c/c was Richard Bramley. Note the interesting mission markers which are five 8 balls with swastikas hanging below. *Holmes*

The F-5E-2-LO, serial 43-28311, was piloted by Lt. Clark who was assigned to the 7th Photo Recon Squadron. The aircraft was adorned with this lovely lady and carried the number 311. It was crewed by c/c Harold Green and asst. c/c Morris Rosenfeld. The ship was lost 12 March 1944 when it crashed into the canal near Vallenciennes, France. *Holmes*

*WRANGLER*, a P-38, was assigned to the 20th Fighter Group, also known as the "Loco Group" because of its numerous and successful attacks on locomotives. *Ilfrey*

*Sweet Bets*, a P-40, was piloted by Lt. Charles Jaslow of the 87th Fighter Squadron, 79th Fighter Group, which was stationed the Western Desert of North Africa. The aircraft carried the squadron insignia of the 87th, the *Skeeters*. On the tail was a picture of the Disney character, *Dopey*. A close up of the nose of *Sweet Bets* reveals the machine-gun-carrying Skeeter which was the insignia of the 87th Fighter Squadron. Lt. Jaslow and the tail art on his P-40, *Dopey* of Snow White and the Seven Dwarfs fame. Lt. Jaslow said he was considered the runt of the squadron because of his size and therefore he adopted Dopey. Dopey was drawn and partially colored by Lt. Jaslow. He colored it with crayons which was all he had available to him. *Jaslow*

*"Jinx"*, is the among the P-40s of the 87th Fighter Squadron lined up awaiting "start engines" order for a mission from Castel Benito Airfield, Tripoli, March 1943. *Jaslow*

Cockpit of later model P-40. Someone thought pilots should have a "side vent" to throw out cigarettes or cigars so later models had that side window which could unlatch without opening the canopy. Note the stylized gold pilot's wings with the pilot's name. The kill mark is for a Bf 109 Lt. Jaslow shot down 2 April 1943 in the Wadi Akarit area. *Jaslow*

Making up "B" Flight of the 87th Fighter Squadron, 79th Bomb Group, after their transfer from the 57th Fighter Group prior to deployment overseas are : Scotty Rogers, Frank Huff, James Hundley, Leo Berinati, Ed Holston, John Dzamba, Charles Jaslow, and Red Crossley. They are seen here with a P-40 belonging to the Skeeters. *Jaslow*

*SHILLELE*, a P-40, carries the insignia of the 65th Fighter Squadron, 57th Fighter Group, a red game cock. This aircraft, belonging to Philip Cochran, is at Bradley Field. Pictured with the aircraft is Charles Jaslow. *Jaslow*

Charles Jaslow stands beside his CO's P-40, *Scrappy,* in this photo taken May 1942 at Bradley Field, Connecticut. Note the yellow and white design on the nose of the aircraft. *Jaslow*

This captured German Focke-Wulf Fw 190A, in American markings, was taken by the 85th Fighter Squadron and sports their squadron insignia "Skulliver", the flying skull. The aircraft was repaired to flying condition and flown in Italy by members of the 79th Fighter Group both for sport and for familiarization. The Fw 190 was destined to become one of the most formidable fighters in the Second World War. This photo shows the wide insignia yellow undersurface and band around the fuselage common to captured enemy aircraft which were flown by the allies. *Jaslow*

Another of the German aircraft captured by the 79th Fighter Group was this Messerschmitt Bf 109. Lt. Charles Jaslow sits in the cockpit of the Bf 109 dubbed *Irmgarde*. Lee Gossick, called "Goose" by his friends, commander of the 79th Desert Rats used it to instruct in the combat characteristics and performance of the 109 to fellow pilots of the 79th Fighter Group. "Goose" Gossick who was assigned to the 87th Fighter Squadron, retired from the U.S.A.F. as a Major General. During his service in the 79th Gossick destroyed two 109s and damaged three. *Jaslow*

*Sweet Sue*, a P-47D Thunderbolt, #67, serial 228143, was assigned to the 89th Fighter Squadron of the 80th Fighter Group. The pilot pictured here is Lt. Len Boyd taken in 1945 at Dhudkundi AFB, India. *Boyd*

STANLEY'S STEAMER, a P-47N, was piloted by "Ace" Lt. Stanley J. Lustic, who was assigned to the 19th Fighter Squadron, 318th Fighter Group on Ie Shima. *Taylor*

*BOTTOMS UP*, a P-47N, serial 44-87962, was assigned to the 19th Fighter Squadron, 318th Fighter Group. Its pilot was Lt. William H. Mathis. *Taylor*

*DRINK'N SISTER*, a P-47N, serial 44-47319, was piloted by Capt. John E. Vogt of the 19th Fighter Squadron, 318th Fighter Group. Capt. Vogt is credited with 5 kills making him another of the 318th's aces. *Taylor*

*RED-E RUTH*, a P-47N, belonged to the 19th Fighter Squadron, 318th Fighter Group. Under the cockpit is the insignia of the 19th Squadron, a game cock prepared for fighting. It has the blue spinner, cowl flaps and cowl nose . The nose is edged in black. *Taylor*

*ARIZONA LORAINE*, a P-47, serial 43-3250, was piloted by Earl L. Kielgass, of the 396th Fighter Squadron, 368th Fighter Group. This color was captured as part of a 16mm sequence taken during November 1944 in Belgium. The art was the design of Earl and was a depiction of the Arizona desert complete with cactus, cow skull, and mountains. Like many pilots, Earl incorporated his wife's name into the art. *Kielgass*

*"OLE COCK"* was the name given to this P-47 belonging to Capt. Donovan Smith of the 56th Fighter Group. *Hess*

This art was on the P-47, serial 42-7938, flown by Lt.Col. David Schilling, named *Hairless Joe*. Schilling scored 22.5 aerial victories during his assignment to the 56th Fighter Group. All of these were in the Thunderbolt. *Hess*

The name given to the P-51, *MY ACHING ASS*, assigned to the 72nd Fighter Squadron, 21st Fighter Group, can be attributed to the pain of sitting in the confines of the P-51 during long and tiring eight hour missions. *Stauffer via Copeland*

*Fertile Myrtle*, a P-51, serial 44-63975, has the yellow spinner of the 72nd Fighter Squadron, 21st Fighter Group. *Stauffer via Copeland*

The F-51D, *Kitten* was piloted by Cap. Charles McGee of the famed Tuskegee Airmen. McGee named the P-51 *Kitten*, not in honor of a female, but instead in reference to the smooth running Rolls Royce Merlin engine. "It purred like a kitten." explained McGee who flew both for the 332nd Fighter Group during World War II and later for the 67th Fighter Group in Korea. Pictured here we see the aircraft loaded with air to ground rockets and long range fuel drop tanks. *Spry*

信念의 鳥人!

This P-51 was flown by Col. Dean Hess, who flew P-51s in the Korean conflict as well as World War II. Hess became well known for his humanitarian spirit. During the Korean War Col. Hess, an ordained minister and fighter pilot, saw to the rescue of hundreds of orphaned Korean children in what was to become known as operation "Kiddy Car." Pictured here we see Col. Hess flanked by two R.O.K. (Republic of Korea) pilots. The Colonel's P-51 is seen in background being serviced. *BACON, via JAMES*

THE
GREAT SPECKLED
BIRD

*THE GREAT SPECKLED BIRD*, a P-61, was piloted by Earl Dickey of the 416th Night Fighter Squadron. This photo was taken in Horsching, Austria. *Thompson via Pape*

Dusty

A panther on the prowl is seen on the P-70, *Dusty*, flown by Fred Secord of the 6th Night Fighter Squadron. This is possibly an unofficial squadron insignia. The photo was taken in 1943 before the 6th received their P-61s. *Thompson via Pape*

*PUDGY III* was piloted by Maj. Thomas B. McGuire, Jr. of Satan's Angels fame. McGuire was the C/O of the 431st Fighter Squadron, 475th Fighter Group. He was credited with thirty-eight kills before his untimely death 7 January 1945 during his service with the 475th Fighter Group. *Hernandez*

*Our Lil' Lass*, an F-82G, was piloted by 1st Lt. B.J. Buckhout during the Korean War. Note the Korean Occupation ribbon on the base of the canopy. *Spry*

Jim Crumley leans out of the cockpit window of the P-61, *HUSSLIN' HUSSY*, of the 6th Night Fighter Squadron. This photo was taken on Saipan. *Thompson via Pape*

*OH'KAY BABY*, an RF-51, was assigned to the 45th Tactical Recon Squadron, of the Oklahoma Air National Guard. As pictured in Korea. Note the polka-dot design on the spinner, wing tips and the tips of the horizontal slabs. *James*

# Index